The Gods Of Science

by

Abraham Kiggundu

Bloomington, IN  Milton Keynes, UK
authorHOUSE

AuthorHouse™
1663 Liberty Drive, Suite 200
Bloomington, IN 47403
www.authorhouse.com
Phone: 1-800-839-8640

AuthorHouse™ UK Ltd.
500 Avebury Boulevard
Central Milton Keynes, MK9 2BE
www.authorhouse.co.uk
Phone: 08001974150

First published by AuthorHouse 9/12/2006

ISBN: 1-4259-3725-X (sc)

Printed in the United States of America
Bloomington, Indiana

This book is printed on acid-free paper.

*The cover photo was taken on Easter Island and has been used with
the permission of Dr. Georgia Lee and the Bradshaw Foundation.*

To all those friends, peers and family with whom I have discussed the contents of this book over the years. Without your feedback this would not have been possible. I dedicate this work to my two sons, Isaac and Benjamin whose bright smiles continue to provide a light onto my journey through life.

Contents

The Gods Of Science

1   Introduction

From the dawn of time, man has been engaged in what has seemed to be an insurmountable battle. The battle to understand and hopefully overcome the constraints of his existence. As cliché as those first two sentences sound, dig in a bit deeper and you'll be well on your way towards appreciating the very foundations of reality. You will have taken one huge stride on a journey leading to the answer to the biggest question of them all. Why?

We have all at one point or another wondered about reality and our existence within it. Many have made the pursuit of answers a life long quest and throughout the centuries, many of those lives have come and gone yet the questions live on. Ever since that day two or so years ago when I began scribbling thoughts on pieces of paper my sole goal has been to prove that the search for answers is not a sacred task bestowed onto the privileged few but a goal to which every one of us can and should make a contribution. In my opinion this grants humanity the great hope that in the grand scheme of things, we are moving in the right direction towards answering these questions, especially in today's changing world.

I believe that somewhere amongst us lies an idea whose contribution to this cause will stand out in the centuries to come, much like Galileo's did and still does. That idea or concept may not necessarily originate from a renowned scientist or a source of great power and influence within the scientific institution. That source could be anywhere and anyone not realising that those thoughts ringing through their head have the potential to shape the direction of human civilisation over the next few centuries. One characteristic that will likely set this source apart from the rest is an unrelenting urge to ask questions regardless of whether or not they have eminent answers.

I would like to hope that this book will help flush out new ideas by encouraging healthy and fresh debate on what we know about the world that surrounds us and the reality of which we are part. By sharing my experiences in searching for explanations and answers over the last two years I hope to help open up the topics to a wider audience in appreciation of the fact that one does not have to be a decorated physicist to contribute to the truth and help us understand the universe of which we are part.

To get the best out of this book I suggest you read it with the mindset of a child. Approach the chapters with an open and pragmatic yet inquisitive mind and you should come out the other end ready to make an objective contribution to the cause. The book does not claim to be the window into that room that holds the absolute truth we all crave at one point or another in our lives; it simply intends to set the scene for such windows to be discovered.

2 Why are we here?

I thought I'd begin this goliath of a book with what is arguably the easiest question to ask and ironically the most difficult to answer if at all it can be answered in the first place. Everyone irrespective of colour, creed, age, sex and intellect will at one point in their lives, even if implicitly, ask this question.

The question it's self is like a thick impenetrable forest. We all at some point in our lives look at it wondering what unknown treasures and wonders it holds. Some of us are so teased by it that we venture in. After a short distance climbing through the overgrown, dense and unrelenting undergrowth we follow our tracks back. We are welcomed back by cheeky grins from those that could not be bothered and together find more interesting things to do like resurrect a 1970's sports car or make a load of money on EBay.

Some of us prefer to sit outside the forest, glaring at its mystery. Every now and then we pick up an attractive leaf, stone or creature from it's outskirts and spend our life incorporating it into our beliefs about what the rest of the forest holds. Of course, among us, anyone's view or story is as good as another's but we still find it worthwhile to argue about which is right and the winner is more often the loudest rather than the closest to the truth.

Others, in a glowing personification of Man's inherent arrogance, wielding tools from far and wide, big and small, whatever is within their means from big budget goliaths atomic colliders and space crafts to telescopes and microscopes, attack the undergrowth making what appears to be phenomenal headway. Their paths ramify throughout the forest each followed by a stream of inquisitive individuals always a step behind, much like a fan club. At some point, some give up the endeavour and turn back. Other's follow the steam rollers and discover islands of beauty and intrigue over which they spend the rest of their lives musing. Some in fact follow the wrong machine which eventually breaks down in the middle of nowhere and faced with the prospect of spending the rest of their lives back tracking decide to justify and defend their situation venomously to their death.

Every now and then a group or even an individual veers of the beaten path and makes a discovery that others can use to better understand the forest. With this they can exponentially increase their rate of progress, however the more such events occur the bigger the forest appears to become. Their egos take a severe beating and those that do not wave the white flag at this might die heroes and icons if they are lucky or otherwise as ignored entities in straight jackets locked away from society for fear they shall dishearten others in their endeavours.

2.1 The purpose of life

For what reason does life exist?

Surely it does not exist solely to wander hopelessly through some forest. Surely there is a grand purpose even if it lies beyond our comprehension?

To answer these questions let's ask a more obvious question, what is the purpose of a knife?

A knife is built for one purpose only and that is to cut things. Throughout its life as a knife it will either be displayed in all its sharp and shiny cutting glory or used to cut one thing or another. At some point it might even digress and be used to sharpen other knives or even to open screws. The point here is that the creator of the knife had a specific purpose in mind for it which is evidenced by the characteristics of its construction. Throughout its life it might find new uses for its self which will inspire the creator when forging new knives which consequently will gradually acquire drastically different or more versatile designs and purposes.

Humans, and in fact all living things both flora and fauna, are infallibly built with one very specific instinct, Self Preservation and one very dominant ability, the ability to procreate. All known behaviours and characteristics of the living world can be explained along a spectrum that stretches from self preservation on the one end to procreation on the other. Like the knife's sharp edge demonstrates its purpose, man's self preservation and procreation characteristics provide a foundation on which to discuss the purpose of life.

2.1.1 Self Preservation

When you are born to this world your first concern is to ensure your lungs are inflated and providing vital oxygen for the rest of your body to survive in this world. Your next worry is food so you latch onto the bosom of the more voluptuous of your progenitors. Already the signs of one of your two core built in functions are showing.

As you get older society and your surrounding will provide a mould in which to develop this built in characteristic. You will desire the best health advice possible when sick, worry about death, and at the far extreme inhabited by some, even kill others preemptively fearing that their survival in some way endangers yours or their death makes your life more fulfilling.

You are bestowed with 5 senses all of which are constantly feeding you with information predominantly used to preserve your life both in longevity and quality. When faced with objects hurtling towards you, high temperatures, unpleasant whispers, repulsive smells or medicinal tastes, your reflexes kick in to mitigate any threat to your personal wellbeing. The picture becomes even more interlaced as we all in one way or another at some point in life both directly and indirectly also provide protection for others close to us.

At a global scale, we all contribute to the common cause of securing and improving the future of human kind. Science is a critical component of this communal effort because it allows us to better understand the universe of which we are part. With this we are warned of dangers to our continued survival and that of the world that hosts us such as global warming or the likelihood and consequence of asteroids hitting the earth. With advancements in scientific knowledge we are able to improve the ways in which we interact with or utilize the universe. This ability enables us to coerce our environment into better facilitating our continued and healthy existence as well as to plan for any dangers posed to this.

2.1.2   Procreation

The second purpose for which living things are engineered is the ability to bring into existence new living entities of the same kind. Though we are all blessed with the tools to procreate not all of us end up directly bringing other humans into existence. However, indirectly through social structures like taxes, friends, peers and employment we in one way or another contribute to this goal at a communal level. Science's contribution to this objective is to find better ways to preserve and promote the progenitive health of the population at large.

2.1.3   The Mandates of Life

One very interesting point to understand is that in life there are only two things you **have** to do: Live and Die, everything else to differing extents is a choice you make. From the time those two gametes embrace you are living. This state of being is something that is imposed on you; you simply have to live in order to have life. Within the same breath fate also dictates that your life will come to an end and in so doing mandates your death.

You choose to breathe because otherwise you will die and that is against your instinct of self preservation. If you by some emotional trauma overcome this instinct and become suicidal you may choose to stop breathing. The inevitable then happens sooner rather than later. You wake up early and go to work each day because you want to, in one way or another improve the quality of and hence preserve your life. You aspire for greatness, fame or fortune which if achieved will give you and those close to you an edge over others in life. Everything you do in life can be linked back to your inherent instinct of self-preservation and procreation.

Scientific progress is therefore an absolutely critical and all pervading pre-requisite to the existence and future survival of human kind within the constantly changing universe of which we are part. It can be regarded as a mandate of life because it facilitates both procreation and self preservation. In this context, it is imperative that we all hold a personal obligation to contribute to it. Handing such a crucial aspect of our communal survival completely and solely to a select few whom we label as '*Scientists*' is a risk that might well regretfully prove catastrophic for future generations. Each of us should not only harbor an obligation but also feel a responsibility towards actively and constructively contributing to this cause in one way or another. We can not afford to ignorantly allow it to be run down by a minority to the detriment of the rest. Any corruption of our scientific fabric should be dealt with head on and without hesitation or subordinate tolerance.

3 Models, the tools of life

To appreciate science we need to understand the tools that make it possible. This chapter discusses what are arguably the only tools we have towards understanding the universe of which we are part. The entire institution of science and our very perception of reality are based on models. This makes them invaluable in our comprehension of the world around us but also exposes them to abuse and misuse, undoing the very progress they foster. So what are models and why are they so important yet so dangerous to our further understanding of the universe?

3.1 Models and Perception

A model can be otherwise described as a representation. In the context of this discussion, it is used to refer to any representation of the reality of the universe of which we are part. The science of modelling is an outward expression of how we conceptualise our perception of the environment we live in using the ever flowing stream of information coming in through our five senses. In other words, everything we sense through sight, smell, touch, taste and hearing is abstracted into concepts like aesthetic beauty and comparative analysis, which we use to associate, interpret and store this information. Armed with the resulting models we can then attempt to understand and even predict the modelled reality. In the meanwhile, information continues to flow incessantly towards our five senses irrespective of the limitations on the rate at which we can register process and relate it all. We perceive surrounding events a fraction of a second after they occur. In addition, the perception is not formed from the absolute, entire information incident at our senses but rather from samples taken from this at a rate equivalent to that at which we can process the information. Our impression of the world around us is consequently, inherently inaccurate but luckily for us, mother nature arranged things in such a way that this inaccuracy can be ignored for all the practical purposes that allow us to function and relate within the universe. Any less accurate and it would be possible to fall off a cliff before you saw that you were walking towards it, what a different world that would be.

As new information continues to impinge on our five senses we are aware that the world around us is changing and are constantly having to revisit our model of reality. Old assumptions are revised or even invalidated while new ones are formed. Some models provide more perceptive and incisive views into the wilderness than others and by virtue of that are more durable and pertinent to the domain, but that does not make them anything more than exactly what they are; models of reality.

3.2 Is Perception Reality?

By the above definition any adulteration or translation of occurrences in the reality that surrounds us into forms which we can interpret, analyse and hopefully understand will inherently produce models of the reality. The very observation of the occurrences in our universe would not be possible without the use of models. Photons impinge on our retina which are designed to detect variations in there nature and send this information to the brain which consequently constructs a representation of the world around us. It is just a representation and not the complete picture due to the limitations discussed above. This might come as a shock but when we observe a tree that image formed by our brain is not the tree but a perceptual model of that tree. If you could imagine your self reduced to the size of a Gamma ray you would notice that you are able to walk right through that tree as if it were like air. The tree has not changed at all; it is you that has grown smaller and now interact with the world at a different level changing your perceptual realisation of this tree.

It is also possible to imagine that on a grander scale of things, way larger than cosmology is aware of, our universe is impermeable much like we would no longer be able to walk through the tree if we grew from the size of Gamma rays to our normal relative size. This exaggerated illustration provides a vivid highlight of the vast difference between the actual reality of this universe and the perceptual model of the same on which we rely.

3.2.1   The Observer and the Observed

Everything we know and perceive about the universe around us is through models. We have never perceived reality. If this still sounds far fetched then another illustration is in order. When we obtain information about the world around us through our five senses we are either directly interacting with the object of interest or interacting with intermediate objects that have either interacted with or originated from the object of interest. When we interact with an object through touch or taste the process will always results in changes within the subject of the interaction no matter how small. Even touching a concrete wall produces imperceptible changes in the touched wall due to transfer of heat and disturbance of the microscopic structure of the wall's surface. The touch does not sense the wall as it is but senses it in terms of how it physically reacts to the touch. When the hand leaves the wall it leaves it microscopically different to how it found it. In fact, it is arguable that though it may seem the same, the second touch will not sense exactly the same wall as the first.

What the information that travels to our brain is actually telling us is that touching a wall produces a certain sensation which we automatically attribute as a basic characteristic of the wall and can predict that when we next touch the wall we will get the same sensation. Of course for all practical purposes successive touches of the wall are not noticeably affected by the effect of previous touches and the sensation is recorded as part of the tactile model of that wall.

Imagine you were smaller than an ant standing on a grain of sand in the concrete of the wall. This would feel smooth to you and touching the wall grain by grain would lead you to conclude that the wall is smooth. Similarly, a giant capable of holding the planet earth between his fingertips would sense that it is smooth to the touch unknowingly flattening whole cities as he draws his finger across its surface while you and I would continue to appreciate the earth's varied surface profile. In all the above cases the sense of touch is not recording reality but an impression of reality relative to the observer. The sense of taste follows the same argument though the nerve signals it sends to the brain have a chemical origin rather than a physical one.

The other three senses are similar in the way that information is constructed at their receptors and forwarded to the brain in that they need to interact with some physical entity. They differ in that the entity they interact with is not necessarily the entity of interest, it is an intermediate entity and the information collected from its interaction with the sense is secondary. We perceive the primary entity by virtue of a secondary '*conduit of information*' whose perceivable characteristics are either changed by its having interacted with the primary entity or are a result of it having been originated at the primary entity. This secondary level information allows us to perceive objects that are a distance from our current location. The cost we pay is that the further away the primary object the longer it takes this conduit to transport sensory data from this object to our senses and depending on how fast it travels the information will get to our senses after the event thereby introducing a disparity between what we sense the object to be and what it really is at the instant during which we sense it. This situation is made strikingly obvious when we try to observe stars other than our sun the nearest of which , Proxima Centauri, is so far away that it takes light 4.2 years to travel from it to us here on earth. We have gained by being able to sense something far away but we are proportionately loosing in that the information we receive is stale and not representative of the current state of the star.

3.2.2   A double edged sword

Our innate use of models in perception enables us to quantize perceived aspects of our universe into discrete chunks. Space is divided into discrete positions and units of length whereas time is divided into discrete epochs and units of duration. We hit limitations in how small these discrete units can be while still making sense yet we have no reason to believe that they can not be made infinitely smaller. In the end we are forced to concede that the reality we are modelling is not discrete but continuous leaving us with a gap between the limits of discretion reached by our models and the apparent continuity of the reality they represent. Many will devote their lives trying to find ways to reduce and eventually remove this gap forgetting that it is a result of the inaccuracies intrinsic within the models that facilitate our perception of reality. The only way to completely be rid of it is to avoid having these models in the first place. However, these perceptual models are the only way we currently have of understanding reality and the universe at large so in these efforts we end up getting stuck effectively chasing our tail.

Models allow us to compartmentalise a huge quantity of information more readily. For instance, if I was to describe my current location I could say I'm sitting on a chair in the top floor of a house near a road leading to Charlton through Greenwich which is in London an area in the central potion of England which lies close to the North Pole etc. This is a lot of information just to communicate my location to someone across the world which does not even provide an accurate representation of where I am. By using a 3 dimensional system of coordinates based on a point whose position can be accurately determined I can replace all this verbal diarrhoea with 3 numbers: northing, easting and height. As a model these are both more accurate and more concise than the descriptive alternative. This ability to compact and simplify information allows us to store more information about the world around us in our brain. Without this efficiency our brain might be overwhelmed by the information it receives through our senses since it will have to store each and every iota of data.

3.3 Looking Beyond the Models

The only way to perceive reality without models is to be able to sense it directly and instantaneously at any distance, without interfering with it in any way and without translating the information sensed. This would result in an irrationalised and uncontained impression of the universe wholly independent of the observer.

We would be able to see and talk to everyone on planet earth and in the universe simultaneously and instantaneously. In fact, we will instantaneously all of a sudden sense and know the nature and content of the entire sensible universe. We would no longer need to make models out of observations which can help predict the result of more detailed observations since we would simply already sense the result of these more detailed observations. We would have no want for more information since we already instantaneously have all the information that exists. The big question would no longer be what is the universe made of but '*What is the universe capable of?*' Instead of trying to find out what's out there (since there is nothing left to find out) we would instead be more interested in how we can influence the nature and behaviour of the universe and thereby affect or even change reality. The models then would not be related to what things really are because we already have this information, but what things can be changed into. Can the universe be manipulated such that there is no gravity? Or even such that a cat becomes a dog? This is obviously a level well outside the realm of science and our perception of the world today. The important thing to realise is that until we are at that level we will continue to understand and perceive the world around us through models. Consequently we should appreciate not only their uses but also their inherent limitations.

The point in this discussion is that our senses translate information from the world around us and convey it to the brain which relates it all to form a representation (model) of reality. At no time do we actually perceive instantaneous reality. We use the resulting model to understand the universe and make educated assumptions about it which we can use to further test the model. On the flip side, the model we use to make observations of the universe indirectly imposes limitations on our understanding of the same. Our very perception of reality is a model yet it is the most extensive and inclusive impression of reality that we have. By accepting this we can begin to appreciate that there are limitations to how much we can reliably infer from this model without actually perceiving reality.

3.4 Dangers of extrapolation

Through simplification and generalisation of models it is possible for the brain to implicitly record a lot of information about the world around us. For example, I know through experience that if I push an object off the edge of a table it will fall to the floor and produce a sound. Depending on how heavy it is and the nature of the floor, it might cause damage to its self and\or the floor. From this generalisation I can predict what would happen if other objects, (which I have never seen fall off my table onto the floor) were pushed off the edge. I can predict that a balling ball would leave some damage on a wooden floor and an ice-cream would make a mess physically damaging its self rather than the floor. What I now have conceptually in my head is a model of the consequences of an object being pushed off the edge of a table and allowed to fall under gravity onto the floor. My perception of this action is its self a model of the underlying reality so in effect the concept of an object falling off the edge of the table is a model of a model of reality. A model twice removed from the underlying truth.

As we can with any model we can extrapolate from this model to make predictions. This is fine for everyday objects but what if a grain of material from the surface of a neutron star weighing as much as a few trucks was pushed off the table. We know for sure it would behave more like the heaviest of objects we are aware of than anything else but how far into the earth would it fall. What if it was a cup full of this stuff, surely there would be global consequences like changes in the earth's spin or its motion around the sun which would effect climatic changes which might lead to the extinction of our species on earth all just by dropping a cupful of neutron star dust onto the earth. This is all a wild extrapolation from the original model necessitated by the introduction of parameters outside of our day to day experience. The degree of certainty about the outcome has reduced and the number and variety of the alternative outcomes has increased as we stretch the model to its limits. Which begs the question, when is a model no longer adequate to predict the outcome of an unknown experiment? Do we have to get to this point to realise the models shortcomings or is there a way of realising this before?

3.4.1   Lessons from the past

The abuse of models is not something new to humanity. Early philosophers believed that the earth was flat. This model was the basis of the early Greek maps[1] but by the 1st century with the spread of the Roman Empire as the means to travel to far away lands became established, it began to dawn on us that this flat earth model was incorrect[2]. The earth's surface was in fact curved. This was the only way to explain why ships were observed to recede over the horizon, disappearing hull-first. It had taken centuries of advancements in transportation

---

[1] Anaximander (610 BC – 546 BC), who introduced the sundial to Greece and Hecataeus (550 BC – 476 BC) produced some of the earliest maps known. These were based on the assumption that the earth was flat.

[2] It was around 240 BC that Eratosthenes (276 BC – 194 BC) estimated the circumference of the earth but it was Pliny the Elder (23 AD – 79 AD) who popularized the new spherical earth model.

and communication before the flat earth model could be identified as incorrect.

Just as the earth was established to be flat, another doomed model which portrayed the earth as the centre of the universe was formalized[3]. Today we know this too is not true. We also know that atoms are significantly more complicated than the Billiard balls of John Dalton's days. All these were well respected and useful scientific models in their time which were replaced as we got to learn more about the world around us. While these models were in their prime anything to the contrary was laughed at and in the case of Galileo it was a heretical crime for which he was persecuted. Now days, we look back with an air of confidence that we know much more about the universe than we did then. We should also realize that the models some of us hold up as unchallenged truths will most likely be proved inadequate through the centuries to come. The scientific evangelism characteristic of many of today's theories and the institutionalised bias that favours one theory over another bear analogy to the Catholic Church during the 15th century. The scientific institution reveres icons and figureheads like Einstein making it tantamount to professional suicide to formally challenge the model legacy they have left us with. These are the Gods of Science and just like religious gods we can not prove them as absolute truths but

---

[3] Ptolemy (90 AD – 168 AD) proposed the geocentric Ptolemaic system. And it was not until the 15th century AD that Galileo and Copernicus replaced this with the more correct heliocentric model.

hold onto them by faith until future generations render them ridiculous just like we did to the Ptolemaic system. As humanity we need to start appreciating that because models are inherently separate from reality there will always be more than one model that is valid for a given subset of reality. With time the weaker models will be left by the wayside however in their place will arise other more worthy replacements. This will only cease when we hit the limit of human perception and hence can no longer make new observations. However, if our perception of the universe has no limits, then this cycle will cease when we are able to sense the universe directly and instantaneously without the use of models.

3.5 Mathematics as a model

Mathematics enables us to relate and communicate the quantity or value of entities within the universe. With this we are able to quantize the universe well enough to make accurate and meaningful predictions. The underlying principle of mathematics is the very simple yet powerful concept of numbers. However, throughout history this concept has had to undergo various tweaks to continue to support the growing use of mathematics in understanding our universe.

The number zero was introduced to represent the idea of nothingness in mathematics but it also introduced problems within the number system. When we started using it in equations we discovered that dividing an ordinary number by zero makes little sense and results in infinity. The number Infinity is yet another introduction to the number scale whose meaning we are still trying to pin down. It is a number which seems to have the license to break almost all known fundamental rules of mathematics. No matter how many times you multiply it by it's self you still get the same number, Infinity. In equations it can best be translated to mean '*that value which is too big or too small to know*'.

As mathematics was used more and more to model reality we continued to stumble across limitations in it which resulted in the creation of complex numbers and even numbers that can not be represented on a number scale like pi. The success of mathematics as a model lies in our ability to constantly tweak it to fit new observations and situations. However, the fact that it needs to be tweaked every now and then reminds us that it does not provide an absolute representation of reality but remains a mere model of reality.

If introducing a new mathematical concept and numbers somehow leaves the old concepts undisturbed and somehow completes them we feel that the model is getting better and giving us an improved understanding of the reality it represents. As a model, mathematics has deservedly earned its place as one of Science's Gods.

3.6 The Big Bang, a model example

The big bang as a model deserves a book all to it's self. It has grown from a religiously inspired anthropocentric view on the life of the universe into a huge and unwieldy mathematical model which is constantly being tweaked to keep it in line with new discoveries. Of recent, some of these needed tweaks like the variable speed of Light (VSL) theory[4] have began to chip away at established beliefs within the scientific institution as if to signal the need of a fundamental rethink on the validity of these founding theorems.

As observations like the background radiation are made they are neatly and religiously explained through the big bang model giving the impression that it is the only model that can accurately describe the history and foreseeable future of the universe. All this ignores the fact that this is just a model and as such there are other models many of which are much simpler, which can equally describe the universe but which have not been blessed with an equivalent amount of scientific effort and are hence comparatively underdeveloped. Scientist then make the mistake of implying that the big bang is the true picture of the universe laughing at any who challenge this point of view much like the Catholic priest did for the Heliocentric system in the 15th century. The scientific institution's equivalent of the stake in the 15th century is to

---

[4] **"Faster than The Speed of Light"**, João Magueijo

severely stunt a young scientist's career if he or she does not conform.

The elegant model that is the Big Bang is a scientific icon and symbol of mathematical triumph. However, as a model its founding assumption that the universe grew out of a cataclysmic explosion can not be proved beyond doubt because no one has ever witnessed it. Even though the scientific observations fit well within this model, one can not help but remember that before the days of the roman conquest, all known observations fitted well within the belief that the earth was flat. It is easy to imagine that in the year 3000 so much will be known about the history of our universe that humanity will look back at the Big Bang not as the absolute truth that many myopic scientists evangelistically hold it to be today but merely as a stepping stone on the path towards understanding the universe. The only truth is reality its self and we should appreciate that so long as we are looking at reality through models we are inherently separated from the truth of reality and have little ground if any to discount other models as worthy representations of this reality.

4 Change and Perturbation

Everything you see, or otherwise sense around you, everything you constitute of from the sub quark entities to well beyond the perceivable universe is due to one simple thing; change. In order to explore this assertion we must first understand what change is.

4.1 What is Change

Change is certainly a crucial pillar of life and reality as we perceive it. Simply defined it is that which occurs between two Epochs. I use the word Epoch here not just to mean points in Time but with a more generic meaning that encompasses points in space, time, state, temperature or any other nature that encapsulates a change of whatever duration, host or type.

To define the two epochs that describe a change we must identify and perceive the nature that changes between them. We can only detect the two as separate epochs if by the use of our five senses and the tools currently available to us we manage to perceive the change thus defined between them. This has powerful implications on how we perceive and interact with the world around us.

4.1.1   Perception of change

Though we may define two epochs, if we fail to perceive the change that sets these two epochs apart we will fail to distinguish them within the context of the changing entity and be fooled into thinking that they are one in the same epoch.

By way of example Let us consider a nature which describes change through certain perceivable cyclic behaviour or markings and can therefore be used as a scale on which to define two epochs. Such a nature might be the sun, an atomic clock or even a measuring tape. Using the subsequent scale we can only distinguish between these two epochs if they span a change greater than the smallest perceivable division of this scale. A scale is only worth using if one or more of these smallest divisions (accuracy) that it can reliably report fit between these two epochs.

Take the celestial scale provided by the movement of the earth relative to the sun as an example. It would be next to meaningless to use the earth's yearly motion round the Sun as the scale upon which the winner of a 100 meters race is decided. We know from experience that such a race is tight and all the runners usually cross the finish line within 2 seconds of each other. To adequately clock this race you would need to be able to measure the position of the earth along its orbit to within an accuracy of the fraction of the orbit covered by the earth in these 2 seconds:

Accuracy = 2/(seconds in a year)

= 2/(10*6*60*24*365)

= 1/15768000................ 4-i

Now the circumference of the earth's orbit around the sun is 9.4×10^{11}m which implies that we need to be able to know the earth's current position in this cycle within:

(1/15768000) × 9.4×10^{11}m

= 59614m 4-ii

We must therefore also be able to determine the earth's position within its orbit 10 seconds afterwards to within the same accuracy of 59 km. Now we know our calendars lose one day every 4 years due to there inaccuracies which we make up for using a leap year. Therefore the accuracy with which we can measure time using the perceived earth's motion around the sun is given by:

Accuracy = 1/(days in 4 years) = 1/1461 .... 4-iii

This means we can only really ever perceive the earth's current position in its orbit around the sun to within:

(1/1461) × 9.4×10^{11}m

= 643394935 m................ 4-iv

This accuracy is 10,000 times worse than the minimum accuracy we require in order to clock the race as per equation 4-i.

In other words, the smallest units (accuracy) which can be reliably reported by perceiving the position of the earth in its orbit around the sun are 10000 times greater in magnitude than those we require to clock the race. This therefore renders the earth's motion in its orbit around the sun a useless scale on which to judge the winner of a 100m race. In fact, those stubborn enough to try will always only come up with a result in which either everyone wins or everyone looses.

From this it also becomes apparent that change is realised or perceived on a scale based on smaller change. E.g. a year (one change) is realised after accumulating a certain number of days (other smaller changes) which in turn are each realised after accumulating a certain number of minutes (further smaller changes) and so on down to the Planck limit.

But wait... why do we have to stop at the Planck scale, what roadblock awaits us there?

4.2 The Planck What?

In 1900 when Einstein had just graduated his disrespect for authority and reputation for laziness meant that he was unable to get an academic job. He became a technical officer in the Swiss patent office eventually acquiring Swiss citizenship in 1901 as he started work on his PHD unaware that this work would more or less pretty much define the direction of the scientific institution throughout the 19th century.

At the same time in 1900, Max Karl Ernst Ludwig Planck a well respected physicist mixed guess work with physical intuition and a heavy dose of persistence to solve the ultraviolet catastrophe (see chapter 7.1). Even he himself labelled this as "*lucky intuition*" but it was his successor at the university of Berlin, Erwin Schrödinger and later '*Herr Doktor Einstein*' (as he was to become known) that would build on this haphazard beginning to redefine physical science all together.

As the scientific understanding of the world around us progressed it became apparent that there were limits beyond which scientific models could not make sense of reality and the world around us. Today, the smallest entities that make any sense to science in our perceivable universe have come to be named after Professor Max Plank in honour for the work he did in sowing the seed that grew into this revelation.

The Plank length is the smallest measurement of spatial length that has any sensible scientific meaning as a length and is equivalent to 10^{-33}cm (i.e. 10^{-20} times the size of an atomic proton). The Plank time is defined relative to this as the time it would take light (the fastest entity we know of) to cross a distance equal to the Planck length and is 10^{-43} sec. Because this represents the time the fastest entity we know of will take to travel the smallest distance that makes sense, any value of time smaller than this has no sensible meaning within reality as we perceive it. These two definitions provide the foundation for defining the other three Planck entities, namely, the Planck energy, Planck density and the Planck.

4.2.1 The world that lies beyond the Planck

Why does it seem as though the world simply ceases to exist beyond the Planck? In any case, does anything exist beyond this boundary and if so, what? To answer these questions it is important to understand what is meant by a statement such as '*... the smallest length that has any sensible scientific meaning...*'

A length is a type of change which quantifies the spatial separation between two Cartesian points (or epochs). In other words, a spatial or positional change is required for an entity to traverse from a point **A** to a point **B** so long as point **A** and point **B** are at different spatial locations. Length is then a measure of how much spatial\positional change occurs during this traversal. When physicists talk of the smallest length that has any sensible scientific meaning, they are talking about the smallest spatial\positional change that can not necessarily be perceived but can be used in the classical scientific models of spatial reality to derive useful information pertinent to the world we can perceive. If **A** and **B** are separated by less than this length, there is no known means to physically distinguish them as separate points at separate locations. In fact all the tools and mathematical models available to us will wrongly report that **A** and **B** are at the same location. We can stretch this point further by saying that though **A** and **B** are physically at different spatial positions we see **A** both at the position of **B** and at that of **A**. inversely, we also see **B** both at the position of **A** and that of **B**. Once you start thinking like that you are well into the weird and murky world of quantum physics which shall be discussed in chapter 7.

Similarly, a unit of time is a duration between one epoch and another. In other words a <u>change</u> in time is required to move from one epoch to another. By definition the Planck time is the smallest separation between two Epochs beyond which we know of no means available to distinguish between those two epochs. Beyond the plank time we observe that Epoch **A**, though it is not identical to Epoch **B**, occurs at the same instance as Epoch **B**. In fact, again, Epoch **A** occurs at both the instance of Epoch **A** and … yes you guessed it, the instance of epoch **B**. again the argument works vice versa as well just like was shown with the lengths.

It's important to understand here that at the Planck scale, we as perceptive beings in our reality and with the tools we have to hand at the moment are at a limit in our perception of the world and its reality. There is no reason for spatial and time coordinates to not continue on beyond this point and I dare say, indeed they must and do. Any argument that stands contrary stands in the same arena as those ancestors of ours who thought that because the world looked flat as far as they could see then it **was** flat and travelling too far would result in falling off its edge.

There actually is a logical argument that supports the continuation of scales beyond the Planck scale. This starts with an entity which has a spatial extent of two Planck lengths. Once split into two (not necessarily by physically separating the two halves) this entity can be considered as equivalent to two such entities, each a Planck length in spatial extent. By admitting that each one of these two entities are a Planck length wide you are logically admitting that the distance between the mid point of one entity and it's edge is ½ a Planck wide. No one can argue that half a Planck is the same size as a Planck any more than they can argue that the moon and the big toe on my left foot have the same spatial extent (size). This is irrespective of the fact that many of the scientific theories we hold in high esteem today, at this point actually start to imply that the two are practically equal.   The argument continues to an infinity division of the Planck distance and time. So what is the significance of the Planck units? Why are they important?

4.3 Why use the Planck?

The Planck units even in there very definitions reflect a limitation on our ability to perceive the world around us.

Of the information that we collect about the world around us via various conduits and interactions with our five known senses, light is the medium that appears to have the greatest band width. We get more information from light because of two reasons:

- The unit of information within light is smaller than the units within sound (air molecule), smell (molecule), touch (hair follicle) or taste (molecule).

- These small units are as close together as they are small and travel at speeds way above any of the conduits that supply the other indirect senses (i.e. smell and hearing). When sensing the world at a distance this makes the throughput of raw data per square meter per second unimaginably huge. Suffice to say we do not use all this data; our senses simply take samples and generalise.

So light seems to be the best conduit of information we can use to sense detailed information about the world around us. However, at the plank length our senses, whether aided or unaided can no longer distinguish between light originating from one end of this length and light originating from the other end of the length. We can only perceive or deduce that there is light originating from the vicinity of this Planck length even though in reality it is originating from two distinct points at the ends of the Planck length.

If this is a limitation of light (i.e. light is to a Planck length what the earth is to a tennis ball) then we shall have to find a new medium through which to perceive reality in order to see the world beyond this length.

If this is a limitation of our senses then no doubt (especially with the advent of nanotechnology), one day we shall devise aids to our senses that allow us to distinguish the world beyond this scale using light.

Planck units do not in any way suggest that the world beyond is unreachable or non-existent; they simply draw a line at the boundary of the sandbox of our existence. Boundaries only exist to be crossed and hopefully future generations will prove that this one is no exception.

5 Time

"Dost thou love life? Then do not squander time, for that the stuff life is made of."[5]

Time is an every day concept which largely governs and defines our realm of existence. It would be impossible to even attempt to discuss any of the topics in this book without it. In fact, neither this book nor the topics it discusses nor, for that matter, the entirety of the universe of which it is part would exist without time. What is Time at a fundamental level and why is it so crucial for the existence and perpetuity of our universe? How does such a seemingly simple concept define the reality around us? Can time flow backwards and if so, how? All questions that have intrigued humanity for time (sic) immemorial. To attempt a worthy response to these queries we must consider the issue from first principles.

---

[5] Benjamin Franklin, the earliest of the founders of the United States of America.

5.1 What is Time?

It takes me a certain amount of time to drive from home to work. If I get there after a given time I'm late. Time comes at the end of the day when I have to leave work for home and on some days, it does not come soon enough. We all live surrounded by the concept of time but what is it? If we first consider how humanity "*knows*" that time exists we will be in a better position to understand what it essentially is.

All around us wherever we are, we are constantly reminded of the existence of time. Our bodies run biological clocks whose various cycles tell us when it's time to go to sleep, when it's time to wake up, when it's time for puberty and eventually winding down to the time for us to pass away. My watch changes with time, to mark it, and there exists innumerable other natural and man made cyclic events which all mark the existence of time from the sun to atomic clocks. All these changing entities serve to remind us of the passage of time and where possible give us an accurate measure of this. We contextually interpret the perceivable changes going on around us to derive a subjective impression of the existence of epochs and the durations between them. This impression helps us to accumulate a temporal model of existence which we can extrapolate to help accurately describe the past and better predict and hence plan for the future.

From the discussion above it becomes apparent that time is a concept which at the fundamental level is defined by change (see chapter 4). Simply put, without change there is no time. This is a statement of fact and not an abstract concept or theory. By way of example, imagine that by some will of nature everything in the universe we know of literally froze where it is. Imagine everything from the cosmological motion and characteristics of galaxies to the motion and characteristics of photons all the way down to, and beyond the Quarks and Gluons that are said to constitute all matter, stopped changing in both position and all other nature relative to each other. Photons from the markings on the ruler measuring the distance between two points would be stuck in mid air between the ruler and the observer. Any photons that manage to make it into the eye before this freeze and are in contact with the retina at the point of the freeze would be useless since there would be no signal being generated at the retina much less travelling from the retina to the brain as we know it. All the chemical and dynamic processes in the brain its self that help it perceive the world around us would be frozen. All that will remain outside of this freeze is that which is not limited by our perception of change and hence time. It's interesting to note that this might actually be the ultimate test for the existence of a *'spiritual soul'* which by definition transcends the concept of change as we know it and in so doing it is not constrained by time. This soul would simultaneously exist both for eternity and for an infinitesimally short time, both of which are durations of time we can not perceive. In this hypothetically frozen world, how would time be perceived? The answer is that there would be no time to perceive in so far as we understand and define it in our reality. In other words, without change our concept of time is no longer valid and ceases to exist.

A world like that above, without change, deprives us of the means with which we measure and mark the concept of time. The earth will never move relative to the other planets or the sun, stars will not twinkle, the quartz crystals indispensable in the modern day measurement of time will not oscillate and even the atomic clocks will freeze showing no perceivable change at all. In this world we are lost and can not even imagine how we would define time. A year, a day, an hour and even a nanosecond would all last for an infinite and immeasurable duration simply because no change is available to provide a scale on which time can be measured. As such, we can not define the concept of time in this imaginary world and for all practical purposes, it suits us well to assume that time would no longer exist. The only way out of this tight corner is to define time in terms of something beyond change which would still be able to mark time in the frozen world described above. In other words, the definition of time as change stands at the boundary between the universe as we know or perceive it today and the unknown. The definition is adequate for our realm of existence but by no means absolute. However any valid extrapolation of this definition of time beyond the boundary will have to explain how time would be defined in the changeless world imagined above. If an absolute definition that transcends the concept of change is at all possible it lies well out of our reach, at least for the foreseeable future. Until future generations prove otherwise we have no other choice but to stick with the definition of time through change.

5.2 *Playing with Time*

The laws of physics as understood today do not forbid genuine time travel. In fact the mathematics of Einstein's theory of general relativity allow for the existence of tunnel like short cuts or '*worm holes*' through both space and time. Through the years a number of thought experiments have been proposed which challenged the ingenuity of some of the greatest physicists of the 1900's regarding the implications of this area of science. In a way commanding that either today's physics should be discarded and thought out afresh or simply fixed to logically explain its tolerance for such manipulation of time. With a better understanding of the concept of time we are well equipped to discuss how, if at all it can be manipulated.

5.2.1   Worm holes and billiard balls

To illustrate inconsistencies with today's concept of time travel especially using wormholes as gateways, the billiard ball problem was proposed. In the problem a billiard ball is hurled into a worm hole connected to the past at an epoch moments before it enters the hole. On coming out the other end of this worm hole the ball collides with its original self before this original self enters the hole knocking it away from the entrance. Consequently, the original billiard ball never enters the hole and hence never emerges from the exit to knock its self away. But none of this would have happened if the Billiard ball did not enter the worm hole in the first place. Illogical and self contradictory as this might seem, mathematical physics did come up with a number of proposals for a solution to this problem. Some conjured up multiple duplicate and exclusive dimensions of existence but the most tolerated proposal borrows from the world of Quantum Physics, which ironically, it's self has multiple interpretations some of which again conjure up multiple dimensions. It was through the work of Richard Feynman and more recently, Thorne[6] and Navikov[7] that the quantum Physical proposed solution was consolidated. It however simply pushes the issue over into quantum mechanics by saying that the ball is in a '*superposition of states*' (see chapter 7.3.3). In other words the same ball both goes through the tunnel and **does not** go through the tunnel. It is only when it is observed that it collapses into one state or the other. As discussed in chapter 7, there is much more to the weird '*brotherhood'* of quantum physics than meets the eye raising valid questions against this as

---

[6] **"How wormholes can act as time machines."**, M.S. Morris, K.S. Thorne, and U. Yurtsever, PRL, v.61, p.1446 (1989)

[7] **"Inelastic Billiard Ball in a Space-time with a Time Machine."**, Mikheeva and Novikov, Physical Review D v47, n4 p1432 15-Feb-1993

a solution to the problem. Ironically, Thorn suggests that in principle it should be possible to carry out measurements on the ball both before entering the tunnel and afterwards to detect that it actually has travelled through time yet quantum physics, knowing nothing for sure, can only provide a statistical degree of certainty and not a surety that this has happened (see chapter 7.3.7).

All this scientific backing has since been fuelling the entertaining imagination of science fiction writers but what would happen if we considered the concept afresh and from first principles? If in so doing we fail to derive a clear definition of time which too permits the concept of time travel then we are implying that physics as we know it from the legacy left to us by Einstein, Feynman and colleagues is at best an inaccurate model of reality. However, if we manage to derive a clear definition that also permits time travel then not only do we put a tick against our understanding of the world around us as described by Standard Theory[8] but through this new window on the universe we may be able to gaze out onto a vista that leads towards the eventual realisation of a physical and actual time machine.

_____

[8] The collection of hypotheses presented over the centuries which have been considered successful by the scientific institution. This includes Newtonian gravity, special relativity and Black Holes among others. Some scientist would even go so far as to say that these hypotheses have been proved beyond doubt.

5.2.2   The Crossroads

If we accept that time, in the universe as we know it today is an abstract concept fundamentally defined by change we can then move on to look at what it means to move forward and backwards in time. This principle allows us to limit our thinking and reasoning to the reality of which we are aware and in which we are able to measure the existence of time. In so doing we leave the door open for a methodical and systematic review of our interpretation of time as we continue to discover more about the universe of which we are part.

If we decide that this fundamental definition is insufficient and seek an absolute, extra-universal, extra-change definition then that new definition must continue to define time in the frozen world described in chapter 5.1. Today, we have no means of proving that domain's outside of our universe exist much less that they can be perceived by us. In fact any candidate for such a domain if perceived can in theory, subsequently be incorporated into our current understanding of the universe. This means that anything we can perceive is implicitly of our universe and we can never have knowledge of anything outside of this (if such knowledge exists). If we start to rely on fantasy speculations of domains in other exotic dimensions which we implicitly or by definition can never have perceptual knowledge of we are effectively taking a huge step back to a point where we must admit that we do not understand the fundamental concept of time much less what it means to move through it on this absolute scale. In this realm of science the imagination runs free and exotic theories come *'a dime a dozen'* to those with the ingenuity to loosely connect them back into our universe. It does us little scientifically productive good to speculate about time in this realm unless we can convincingly prove that both the realm exists and that in it, time is no longer encapsulated by the concept of change which defines it in our realm. Unless we can do this there will be no way of both validating the exotic concepts proposed and using them to explain observations in our universe. In addition, these exotic theories need to be concretely substantiated by observations made within our universe otherwise, what observations if any would you make to test them? The concept of observation and experiment as we know it depends on change as described in chapter 4. Surely if it were possible to make observations in another, separate universe, it too should be based on the same concept and can hence be considered part of our own for all practical purposes. The alternative is that our concept of experiment and observation would need to be revised within this new universe if we are to have any hope of understanding it. In other words, what would we mean by observation and experiment within this new universe because it most certainly would not be what we understand it to be in our universe.

Given these two alternatives, until the questions posed by the later can be convincingly responded to it seems more sensible to start our discussion on the manipulation of time from the basic principle that it is an abstract concept defined by change. In this context moving through time is synonymous with observing change upon change within the known universe. The world around us is constantly changing thereby allowing us to observe time passing by all around us. Our perception of time is dependant on which of these myriad changing entities we choose to mark or measure it. By example, consider two quartz crystal watches, one significantly slower than the other such that it adds 2 minutes onto every hour. By this slower watch each day will be measured to last for 48 minutes longer however, we know each day is defined by a period of sunlight (day) followed by a period of darkness (night). After about 15 days the slow watch's reading will start to contradict this natural cycle displaying a daytime reading during the night and vice versa. We shall be forced to resolve the conflict by reconciling the two and the easiest way to do this would be to re-align the watch to the observed diurnal pattern of the sun just like we realign years to coincide with the seasons by adding an extra day every 4 years.

5.2.3   Corrupting the rate of progress of time

With this background it becomes apparent that moving forwards in time is the natural way of things in our changing universe. This forward movement represents the accumulation of change upon change. Conversely, moving backwards in time implies the removal of change from the complex and interrelated state of the known universe to make it as if the change had never occurred.

Let us imagine that the instrument used to measure change and hence time is somehow forced to register more change during the period in which it usually registers a unit change. Only if each unit on this watch continues to be **perceived** as the same unit of time within our universe can it be inferred that time is running in fast-forward mode. Note that this scenario is very different from what happens when you gradually turn the dial of your watch such that the small hand transitions from one hour mark to the next in the space of one seconds. In this case a second as described by the fast running watch is not the same as a traditional second as perceived through the other undisturbed changing entities in the universe. It is substantially smaller i.e. 1 second on the watch is a $(24\times3600)^{th}$ of a second as observed through other changes within our Universe.

Though only one second has passed by, your watch's reading will show that one hour has elapsed. Of course we all know that this is because of your watch's reading is no longer corresponding with the actual passage of time as measured by the rest of the universe?

The point being illustrated here is that the changes that occur in our universe from the motion of Galaxies to the biochemical cycles within our bodies all bear a relation to each other whose respective, relative deviation from each other can be detected and predicted. When our body clocks fall out of sync with the day-night cycle after a long-haul flight to the other side of the world they report the deviation through jet lag. We consequently spend the next three or four days adjusting to account for these relative changes. Artificial change scales like the atomic clocks, intended for higher accuracy measurement of time are inherently more sensitive to such variations and must constantly be adjusted so as to keep them as accurately in line with perceived natural changes as possible. In this way we continue to get the impression that these scales represent and further refine the naturally occurring scales like our body clock and the orbital motion of the earth. By doing this we are implicitly stating that the physical definition of time is the rationalisation of as much of the natural, disparate change that occurs in the universe as possible onto a discrete, detailed mathematical scale whose units and their rate of occurrence appear constant for as far back and as far forward as we require its use. In other words, we are defining the concept of time not in terms of change as it actually occurs in its various natures and magnitudes within reality but in terms of change as it relates to our existence. This modern day physical concept of time implies that change throughout the universe revolves around humanity's interaction with and perception of this universe.

Galileo was sentenced to life in prison in 1633 for publishing the book '*Dialogue Concerning the Two Chief World Systems*' which indirectly promoted the then heretical idea that the earth was not the centre of the universe and revolved around the sun. One wonders what opinion he would hold towards today's anthropocentric perception and realisation of the concept of time.

With this in mind, it becomes apparent that time as a non-tangible concept can not be fast-forwarded. We define and realise time through the change that occurs all around us. What would happen if all the change in the universe both artificially created by man and naturally existing were to proportionately occur more rapidly? The moon would revolve slightly faster around the earth which would in turn spin proportionately faster about its axis. Cars would move proportionately faster on roads and our metabolism would have to proportionately speed up as we move faster about this fast-forwarding universe. On an absolute scale and definition of time outside of this changing universe and unaffected by change we would seem to be living shorter lives. To us within this universe nothing would appear abnormal or different since everything within and around us still behaves the same, relative to each other. We would have no means to detect whether time is moving faster or not from within this fast-forwarding universe and everything will appear to continue as normal. In other words, from a change perspective, if we can somehow find a means of fast-forwarding time then so long as it is done at the fundamental level equally across the entire universe the net effect will be zero.

The notion that a section of our universe can somehow be excluded from this fundamental fast-forward of time is a self contradiction because by the very fact that it is a part of the universe as we know it today this universal fast-forward will not spare it. By way of illustration, lets assume that future generations improve our understanding of time and hence cross the boundary beyond the concept of change enabling the derivation a logical concept of time that would hold in a world devoid of change. They then find a way of realising this concept within any section of our universe such that this section is now governed by the new, more fundamental measure of time which lies beyond the concept of change. Again let us wonder what will happen when all change in our universe is fast-forwarded.

Suppose an artificial measure of time like a wrist watch is placed in this section of the universe governed by the new '*beyond change*' concept. Change as we know it will continue to occur at the fast-forward rate for this watch however the new concept of Time which lies outside of change will not be affected allowing the detection and realisation that time, as defined by change, is fast-forwarding. In effect this new concept would enable us to detect that the universe is changing proportionately faster than before. Without this new concept, everything that defines our universe will change along with the fast-forward and as such we are none the wiser about the fact that Time is being fast-forwarded.

If the watch falls behind slightly as a consequence of its being in this bubble we will simply make adjustments and corrections to its reading as we do when faced with anomalies in the relationships between other changing entities and ignore the fact that this might be due to time it's self behaving differently. It is hence pointless to attempt to detect anomalies in the concept of time through anomalies in a changing entity because all changing entities due to the very nature of change in our universe exhibit cyclic aberrations both in themselves and relative to each other. These aberrations are further randomised as the changing entities interact with and hence influence each other.

What if a naturally occurring measure of time like the sun is placed in this futuristic section of this fast-forwarding universe where the concept of time lies beyond change? If it starts to appear as if it is slowing down we can not simply buy a new sun nor adjust its motion as we might do with the watch as described above. We are restricted to continually making corrections and adjustments to our concept of time as defined by it. In this way, the sun continues to fit into the rest of our fast-forwarding universe. This deviation would be explained away as an unforeseen anomaly and before long a multitude of speculative theories would crop up to try and explain it in the hope of predicting future anomalies. When a day is added to each leap year to keep the seasons in sync with the year no one stops to think that: "*Hold on, maybe Time it's self is behaving differently in the realm of the motion of the earth round the sun than it is in the realm of the earth spinning round its axis*". If we had thought that way we would appreciate that the fundamental reality of time does not dictate that the span of a year should coincide with the seasons and we would consequently not add the leap year in the first place.

Without the constant adjustment and correction of time as defined around us onto a uniform, conceptual and anthropocentric scale, planning far ahead or looking far back into the past would be a lot more difficult. A year would no longer be an integral multiple of whole days requiring any one that needs to convert between the two on a regular basis to walk around with calculators and almanac tables. Things are a lot easier when we impose integral relational order on the natural time scales.

It seems impossible to imagine us being able to detect anomalies within absolute time from within this universe. Corrections like the leap year and the leap second are a constant reminder that we are un-naturally creating the illusion of order and relations between natural entities as we attempt to rationalise reality in our minds. This rationalisation of reality gives us a false sense of security in our knowledge of it. We feel superior because we can now easily predict that July 23rd 2456 will be a summer's day, but will it **really** be July 23rd 2456 on that day? The universe is naturally chaotic and dictating any sense of order on it to facilitate our existence can only ever be a short term and vulnerable solution. We constantly have to adjust bits here and twist knobs there to artificially create the illusion of uniform, cyclic and predictable nature which is far from the chaotic reality. Nature turns out not to be a harmonic, predictable synchrony of change and our insistence on not accepting this could in the end unfortunately serve as a limitation to how much we can understand the world around us. Without an absolute scale of time outside of this universe which is unaffected by change as perceived in this universe we are hence not able to detect changes in the rate of passage of time.

Rewinding time is even more difficult to achieve using our current conceptual perception of time as change. This is due to the fact that we know what has happened in the past and consequently must happen again when we rewind time. To illustrate, imagine waking up, having breakfast and on your way out the door time starts to run backwards. In our perception of time as fundamentally defined by all the interrelated change in the universe, the entire universe would have to start retracing its changes in order to comprehensively achieve this. For your part, you would walk backwards, sit back on your breakfast chair then slowly, spoon by spoon appear to vomit up a perfect, crisp bowl of cereal after which the milk jug would seem to unnaturally suck up all the milk from the bowl and the cereal flakes would inexplicably jump back into the upside-down cereal box. Every moment should occur exactly like it did the first time albeit backwards. You should not realise that time is going backwards because that would contradict the concept of reversal of time. This is because you were not aware of any such reversal when you first went through those actions within normal time. In fact, your brain cells would shift into reverse gear regurgitating all the memory they have stored about the future leaving you ignorant of it having even occurred. You would therefore not even remember having travelled back in time. The only way around all of this is to develop a concept of time that transcends the fundamental definition through change that we have today.

By using multiple dimensions or Quantum physics we can resort to weird science in an attempt to rationalise these weird concepts such that they can be mathematically manipulated. Taking these scientific proposals as unchallengeable truths has analogies with the devout religious person praying to an imaginary God never before seen but believed to exist. Our tendency towards rationalisation within our universe of such a non-tangible concept as time will always end up in scientifically inspired creative imagination rather than fact. All unless we can extend our conceptual definition of time beyond change.

5.2.4   Building a Time Machine

A Time Machine, by definition, facilitates the control and manipulation of time. We can use it to jump forwards and backwards in time to any epoch we choose. The ability to interfere with past events and predict future events with absolute certainty grants unfathomable power but as explained above, time is an abstract concept that encapsulates the accumulation of change in our universe. We currently can not articulate the existence of time outside the boundaries of change in the universe as we know it. It therefore follows that the machine will in actual fact be manipulating change. In such it seems more appropriate to refer to it as a '*Change Machine*'.

If the definition of time is fundamentally based on the concept of change then it follows that the direction of the arrow of time is governed by the accumulation of change. As change piles up on top of change (e.g. as the days go by) we can say that the arrow of time is pointing forwards and in the direction of further accumulation of change. By measuring how much change has accumulated and keeping in mind the duration of a single change relative to the other changes around us (e.g. ageing) it is possible to get an impression of how much time has elapsed. Time is therefore an abstract concept derived from the myriad changes that occur in our universe and how they compare and inter-relate with each other as they accumulate. An interesting point to note is that any change that occurs can be used to measure the passage of time and will be accumulating on the change that occurred before it. Does this mean that the arrow of time points forward even when change is reversed?

We can freeze water into ice as easily as we can melt it back into water. We are able to react hydrogen and Oxygen together to form water and have the power to reverse the process reforming the original gasses. This power does not contradict our perception of time through change because whether we are going from water to ice or ice to water we are observing the accumulation of change on top of change. The change from ice to water happens after and is influenced by the change from water to ice.

In this case the arrow of time continues to point forwards. Now, imagine that we place the cube of ice in a hypothetical '*Change Machine*'. This hypothetical machine would be able to monitor and control the block of ice at a subatomic level, and maybe, with a better understanding of quantum physics, it would be able to monitor and control the cube down to the individual sub quark entities that constitute it. The cube is then allowed to melt, slowly transitioning to water within the Change Machine which all along meticulously monitors all the variables of each and every sub quark entity as these change.

The amount of information that the machine would have to record and store just to represent this simple process is phenomenal. Considering that the dimensions of an atom are of the order of 10^{-10}m, there would be approximately 10^{32} atoms in a cubic centimetre of ice. Each of these atoms constitutes of a number of quantum entities each of which will exhibit some sort of change throughout the process. We know that the Planck second (10^{-43}sec) is the smallest unit of time that makes any physical sense in our reality. The implication of this is that the concept of change which we use to define time in our reality must exist in some form or another at this level in order for us to be able to reliably define this Planck second. It follows that we can expect a Planck entity to undergo at least one '*Planck change*' in a Planck second. Note that this is only the rate of change as approximated at our side of the Planck boundary but as explained in chapter 4.2.1 there is no reason to conclude that there occurs no more than just this one change within a Planck second on the other side of the Planck boundary. However, we are restricting ourselves to that which makes sense within the world as we know it so we shall ignore the world beyond the Planck boundary. Being conservative, we can infer that an atom undergoes at least one change per Planck second accumulating a total of 10^{43} changes in one second. So for all the atoms in one cubic centimetre of ice melting over a period of one second we can very roughly approximate that the number of changes that occur will be of the order of 10^{75} changes. Again take note that although this is a rough estimate, if we had the ability to look over the Planck fence no doubt this number would increase unimaginably. However, we perceive the world within the confines of the Planck so this low guestimate shall have to do for now.

With a bit of optimism we can expect to somehow describe each of these changes using say two computer bytes of information. We would then need 10^{133} Gigabytes of memory just to store one second of information. The number of high spec desk top computers that would be required to store this information is a 1 followed by one hundred and thirty one zeros. To give you an idea of how preposterous this number is; the current world population is of the order of 7 billion. If everyone sat down and built a new high spec 100 gigabyte computer every second of there life from birth, all the way back from the big bang onwards to the expected hypothetical end of the universe we would still not have even half of the computers required to store this mere one second of information. Of course there still exists the problem of acquiring the technology to detect and record all of this in the first place.

With the transition from ice to water complete we can now switch our 'Change Machine' from monitoring mode to manipulation mode such that it works through the recorded information in reverse. Literally undoing all the changes that occurred. Every quantum entity in the pool of water would retrace the recorded changes it went through during the melting process, backwards towards the original block of ice. Now, imagine you could use this machine to repeat the process of record and reverse play back on a whole city. Ignoring the fact that the machine would have to detect record and store a preposterously unimaginable amount of data, you would have created the illusion of localised time travel. As discussed above, this would more accurately be described as 'Change Travel'. Such a machine is still well out of the reach of today's technology.

5.3 Has time travel been proved?

5.3.1 Feynman diagrams and the motion of electrons

In the 1940's Richard Feynman, in characteristic clarity, popularised some simple tools which up until then were only appreciated within the circles that studied Einstein's theory of relativity. These tools are now popularly known as Feynman's diagrams and even today they provide a convenient shorthand notation for a series of (often complex) mathematical equations. They are a code physicists use to talk to one another about their calculations and as such they have proved an invaluable tool towards understanding the complexities and peculiarities within the otherwise prohibitive mathematics of Einstein's theory.

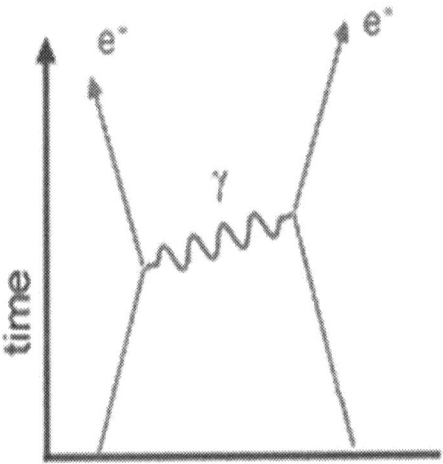

**Figure 5-1: An example
Feynman Diagram.**

In these diagrams (see Figure 5-1) the vertical axis represents time and the horizontal axis represents the instantaneous three dimensional position of a particle. Each arrow thus represents a particle moving through space and time. Vertical arrows represent a particle whose three dimensional position is not changing with the passage of time (a stationary particle). The interesting thing is that as the particle starts to move through space the arrow representing its motion starts to make an angle to the vertical. Really fast particles cover a lot of space (horizontal axis) in very little time so their arrows appear almost parallel to the horizontal axis. Now, according to Einstein's theory of relativity (see chapter 6.2) horizontal arrows represent particles traveling at the speed of light which is thought to be a universal speed limit. In fact, if they could travel any faster the arrow would have no choice but to start pointing downwards of the horizontal. This would imply that the particle would be traveling backwards in time. This observation agrees with the theory of relativity which will be discussed in chapter 6 and the diagrams hence fulfil their job in helping to visualize this theory without resorting to the complex and lengthy equations that describe it.

While playing about with these diagrams in 1949, Feynman discovered that the space-time description they give of a positron[9] moving forwards in time is mathematically equivalent to the description of an electron moving backwards in time.

---

[9] A positron is a short lived particle with a mass equivalent to that of an electron but with a positive charge. It decays to produce an electron and a photon and is in fact the antimatter counterpart of an electron.

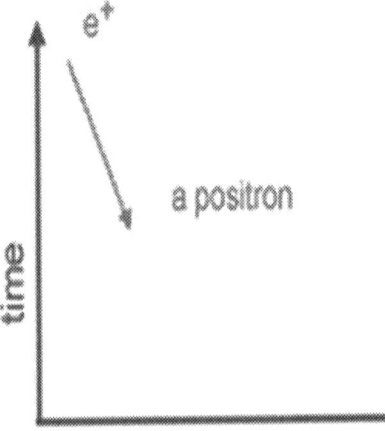

Figure 5-2 representation of a positron on a Feynman diagram

What this meant was that an electron moving through space could meet and absorb a photon to create a positron. If we assume this short lived positron is moving forward in time the situation is mathematically identical to an electron moving backwards in time. We could hence say that the electron absorbed a photon and remained an electron but started moving backwards in time. This statement is still consistent with the complex mathematical model that describes this system.

The positron eventually looses the photon resulting in an electron which we will assume is moving forward in time. This chain of events seems to tell the story of an electron moving backwards and forwards in time as it absorbs and emits photons of light. Does this mean time travel is possible? Surely if the mathematics says it can be done then the rest is a mere formality.

The important thing to understand here is that it is the **model** that is suggesting the notion of time travel and not the reality it represents. Sure, models can be extrapolated to predict characteristics of reality that have not yet been observed but these should be understood within the context and limitations of the model. The model was obtained from a set of real world observations and on playing about with it, it was discovered that some situations it depicts can be easily interpreted as time travel without contradicting the model. This does not mean that these situations are actual representations of real world time travel nor does it imply that time travel is possible. It just means that our model of reality can be interpreted in a way which is not currently observed within this reality but facilitates useful mathematical deductions and predictions to be made from the same model. In fostering these further deductions however, the model has moved further from reality and the fundamental definition of time (see chapter 5) making it even more prone to misinterpretation. We have extrapolated the model to facilitate our mathematical calculations but it is now harder to directly and logically map the model back to certain aspects of the reality from which it came. Any further use of this model to arrive at conclusions about reality should keep this in mind.

What Feynman diagrams actually do say is that as the particle starts moving faster and faster, there comes a point at which it is moving so fast that according to our current limited perception of time and space, no duration of time seems to occur as it traverses the perceived space[10]. This illusion is all down to the limitations in our current perception of time and the world around us. So long as the particle is moving there is change and so long as change is occurring there is a concept of time. If we change our perception of time and base it on the motion of this super-fast particle we realise that time is still passing as measured by the fact that the particle is changing position as it moves through space, albeit the particle can now cover a huge distance in a small duration of time. The horizontal in Feynman's diagrams is therefore not the final frontier towards time travel but represents the limit of human perception[11]. As the particle arrow approaches the horizontal we simply have to find a new and better means of sensing the world and time. If we continue to use light to observe the world, time will appear to run backwards even though no such thing is happening simply because light has become an inadequate conduit of information (see chapter 6.2.3).

---

[10] Technically what has happened is that the epochs of time that define the duration are so close together that, unable to distinguish between them, we perceive them as one epoch and no duration.

[11] In fact, a more consistent interpretation of the diagrams might be that the arrows of faster moving particles are asymptotic to the horizontal and never actually cross it so long as we have an absolute means of measuring time which is not affected by our limitations in perceiving the world around us.

5.3.2   The atomic clock experiment

The possibility of time travel is an exotic inference drawn from Einstein's ground breaking 1906 paper on special relativity as described in chapter 6. It followed from this paper that observing any affects on time due to motion through space would validate this theory. It was this proof that J. C. Hafele and Richard E. Keating were looking for in October, 1971 when they conducted the now famous Hafele-Keating experiment. The published results of the experiment are used by many to argue not only for the validity of Einstein's theory but also for its direct implication that time travel is possible.

The basic idea was to synchronize four specially selected atomic clocks with a standard clock station at Washington D. C. before flying them twice around the world[12] on commercial aero planes. After the flight their readings would be compared with the Washington clock station to observe any discrepancies with this standard that can be attributed to the effect of the journey.

Hafele and Keating however, did not publish the original results in their 1972 paper that pronounced the experiment a successful observation of the *time-dilation* effect predicted by the theory of relativity. Instead they published suspiciously doctored figures which

_____

[12] First eastwards and then westwards to minimise the effect of variations in the earth's magnetic field.

substantiated the theory. An inquiring Dr A. G. Kelly obtained the original, unadulterated results from the United States Naval observatory only to discover that they did not support the results computed in the 1972 paper[13].

The main problem with the experiment concerned the drift rates[14] of the caesium clocks used. This ranged from 2ns\hr to 9ns\hr which resulted in an uncertainty of as much as 300ns in a test supposed to produce a result of 40ns to under 300ns. In other words, any discrepancy between the atomic clocks and the ground station had an accuracy of ±300ns [15] which rendered the measured discrepancy useless in arriving at the figures predicted by the special theory of relativity[16]. In fact, averaging could not make the test more reliable because the clocks were not all equally stable. One clock (serial no 120) seems to have been so poor that according to Dr. A. G. Kelly *"That erratic clock had contributed all of the alteration in time on the Eastward test and 83% on the Westward test, as given in the 1971 report"*. In his opinion

---

[13] View the full account at http://www.anti-relativity.com/hafelekeatingdebunk.htm.

[14] The rate at which the clock looses or gains time due to it's inaccuracies in measuring time.

[15] This figure is arrived at by multiplying the weighted average drift of the clocks by the duration of the journey which was 65.4 hours eastward and 80.3 hours westward.

[16] Einstein's theory predicted a discrepancy of 40ns for the eastwards journey and 275ns for the westward journey.

if this clock had been ignored, the East and West results would have been "*within 5ns and 28ns of zero*". The most stable of the four clocks (serial no 447), by itself was a better experiment than all the clocks together but also indicated zero discrepancy with the ground station. The logical conclusion would be that the experiment did not prove the phenomenon of time-dilation and if the results of clock number 447 are anything to go by, it showed that time does not vary for a moving object and is in fact absolute rather than relative to the observer as described by Einstein. In other words, it would seem that the theory of relativity presents an incorrect depiction of the nature of time. It is however so ingrained in today's institution of science that these contradictions are kept at a low profile while the theory that generated them continues to be promoted as a reference point for most funded, peer reviewed scientific research into the topic.

6 Light

Light has long been a source of mystery and wonder. With its pivotal role in the sustenance of life as we know it, it seems next to impossible to imagine the existence of a world utterly devoid of light. It continues to shape our perception and understanding of the world around us but what is it? Where is it from and why does it exist? Why is it so important to reality as we know it and what would happen without it? Most importantly, what does it actually tell us about the universe of which we are part?

6.1 'Let there be light' [17]

According to the bible light is not only the subject of the very first quote attributed to God but was also God's first creation. A logical inference from this is that God felt that the existence of light was a pre-requisite to any further creation which, if your faith believes in the Bible, could be taken to imply that everything in the world around us

---

[17] The Holy Bible, Genesis chapter 1 verse 3.

including ourselves at the most fundamental level is composed of light. Islam along with many other religions has always tended to refer to light in the context of godliness and truth while darkness seems to always represent ungodly ways and contexts.

Civilisations knew much less by fact about light in ancient times. They observed that if deprived of light, plants would shrivel and die. On realising that the Sun was the source of this light and therefore the protector of all life on earth, they invariably prayed before it as their God. This worship was predominantly fuelled by a fear that without it as a means of appeasement, the Sun God might get angry and deprive them of this life sustaining light. The Aryans[18], the ancient Egyptians, even the Aztecs and Incas of south America all worshiped a Sun God and even built temples to show their appreciation for this God's gift of light. Eclipses were the devil's doing because they prevented Godly light from reaching the people. Much of this spiritual legacy is still evident today, in fact, in India, an eclipse (known as *grahan*) is often followed by scores of people bathing in the holy rivers in order to wash away their sins.

The recorded scientific study of the nature of light goes back to 15BC when Lucretius expanded upon the ideas of earlier atomists by writing that light and heat from the sun were composed of minute particles. There was a lot of theoretical and geometrical analysis of its

---

[18] Ancient people of the Indian subcontinent.

nature in the centuries that followed but the modern views on light did not become formally apparent until the 1500's. Today we know for a fact that light exhibits both particulate and wave like properties, but what is light?

The Electromagnetic spectrum is the range of **known** radiation[19] stretching from the minute Gamma rays to the more commonly known Radio waves (see Figure 6-1). The term light refers to the small potion of this vast spectrum to which our eyes are sensitive.

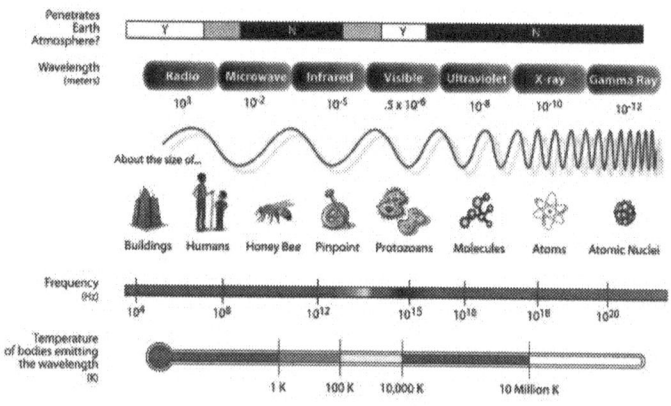

Figure 6-1:
The electromagnetic
spectrum (image: NASA)

---

[19] Radiation is a term used to define a type of energy transfer in which the energy is transferred from the source, through empty space to its destination. The source and destination do not need to be in contact as with conduction nor do they require any known intermediary carrier object as with convection. The term radiation is also used to refer to the actual transfer of energy in this way.

A portion that stretches between the wavelengths of 400nm and 700nm in the electromagnetic spectrum otherwise referred to as the visible potion of the spectrum. It is interesting to note that this portion of the spectrum also happens to encapsulate the smallest known wavelengths of electromagnetic radiation to which the earth's atmosphere is transparent. Is this a coincidence or has our sense of sight deliberately developed to sense in this region in which the detail of small objects can best be observed under natural light from the sun and other extraterrestrial sources? It's hard to resist contemplating that maybe our sense of sight and hence our perception of the world around us would have developed differently if it had developed on a planet where this atmospheric window was further down the spectrum towards the Gamma ray band.

6.2 Analysing Einstein's Thoughts on Light

From the 1600's there have been many ground breaking contributions to our understanding of the nature of this thing called light. From De Broglie to Huygens and Newton and then Young, humanity slowly built up a formalised picture of the physics of light. The picture we have today, at the beginning of the 21st century owes a lot to the writings of a brilliant and up until then largely unknown young man, published in 1905.

Einstein's special theory of relativity defined much of the scientific direction of research around light in the 1900's. It continues to contribute to the foundations of our knowledge of light both mathematically and experimentally. The theory provides among other things a great focal point for any discussion on the nature of light.

6.2.1 Measuring the world around us

Einstein suggested in his '*Special Theory of Relativity*' that the speed of light in a vacuum is constant irrespective of motion on the part of the observer. Those who have studied it in earnest are quick to point out that this is not challenged by the fact that of recent, light has been slowed down by a Bose-Einstein condensate[20] since this light is no longer moving freely in a vacuum.

There are many means through which we can quantify the world around us by measurement. Speed is the ratio of distance travelled over the time taken to travel that distance. From the discussion in chapter 4 we recall that whenever we quote a distance measurement we are referring to the separation between two points and comparing this to an accepted standard unit of separation. The modern day accepted standard or base unit for the measurement of length is the meter and is defined as follows:

_____

[20] A form of matter predicted in 1924 by Satyendera Bose and Albert Einstein which results from cooling certain types of atoms to within a whisker of absolute zero. It has been realised through various experiments in the last decade producing a material that displays unusual characteristics among which are super fluidity, superconductivity and the tendency to climb up the walls of the containing vessel.

*The meter is the length of the path
traveled by light in vacuum during a time
interval of 1/299 792 458 of a second.*

Again, underlying this basic definition is the
assumption made by Einstein's theory of special
relativity that the speed of light in a vacuum is
constant. Einstein's thoughts on light have had a
phenomenally pervasive influence on humanity's
understanding of reality and the world around
us.

Time, yet another means of quantifying the
world around us, is similarly standardised by
comparison to cyclic, predictable phenomena
like the orbits of planets or the oscillations of
crystals. Einstein's theory predicted that time
would behave in strange ways to an observer
travelling at the speed of light. The standards
by which we measure and hence describe the
world around us are by no means universal or
absolute. For example, we have no way of telling
whether the meter as described above does not
change in length with time especially if these
changes in time are experienced by the rest of
the world thereby making it impossible to detect
them from within this universe (see chapter
5.2).

6.2.2   Is the speed of light constant?

According to Einstein's Special Theory of Relativity
the speed of light in a vacuum is constant
irrespective of any motion on the part of the
observer. In other words, two observers moving
at different speeds will always observe light to
travel at the same constant speed irrespective

of their motion. Let us try to understand this statement by way of illustration.

Imagine that two inquisitive cowboys set out to independently measure the speed of a runaway train. One is seated on a hill, smoking his pipe as he watches the train go by while the other is riding his trusty horse alongside the same train. Relative to the two the train will appear to travel at two different respective speeds yet it has only one absolute speed. Reality as we know it dictates that the train can only be travelling at one absolute speed. If it is actually travelling simultaneously at more than one speed then by virtue of the definition of speed, two trains will emerge from one as time passes, one moving faster and separating it's self from the other. I'm in no way saying this can never happen but I think we will all agree that this is significantly inconsistent with reality as we know it today.

The train is definitely travelling faster relative to the stationary cowboy than it is relative to the cowboy on horse back. In other words, if both were to draw a line perpendicular to the train track through each of them respectively as a reference, in any given time period more carriages would cross the line through the stationary cowboy than that through the riding cowboy (whose line continues to move in the direction of the train). The train therefore is perceived to move differently relative to both observers even though it is moving at only one absolute speed which is independent of both observers. There is hence a definite difference between ones **perception** of a moving entity and the **actual** or real motion of that entity.

This is not new science however it provides a foundation from which we can begin to discuss Einstein's constant speed of light assertion. All we need to understand is that the actual speed of any entity is independent of any motion on the part of the observer however the speed as perceived by the observer very much depends on the motion of the observer.

If a beam of light could be observed moving through space, irrespective of how many observers are watching it and how they are moving relative to it, the beam (like the train above) will have a given absolute velocity which we shall assume to be constant and which is unaffected by these observers. The same beam will however be travelling at a different velocity as perceived by each observer respectively, the relative velocity. However, because the speed of light is so great in comparison to the motion of every day observers like you and I, these discrepancies between the perceived relative velocities and the absolute velocity of the beam of light will be too small for the observers to notice. Therefore for all practical purposes they will all observe the same speed of light irrespective of their motion. This does not mean that the actual speed of light is constant irrespective of motion on the part of the observer; it just means that any discrepancies that result are so small they can be ignored for all practical purposes.

To illustrate this further let us first consider the motion of a train which is being observed by an observer who too is in motion such that:

V_{train} = 100km\h,
$V_{observer}$ = 20km\h (in the direction of motion of the train)

The train will be perceived by the observer to be travelling at:

100-20=80km\h............... 6-i

Now, the speed of the train is accurate (indicative of the true speed) to the nearest 1km\h. Therefore the discrepancy between the actual speed of the train and the perceived speed as given by equation 6-I (i.e. 20km\h) is well within the accuracy with which we can measure speed. In other words, though an error of ±1km\h is significant it does not prevent us from noticing the discrepancy of 20km\h above. In fact, for every 100km the train travels, our observer will observe that the train has moved 20km less relative to the observer than it actually has. In general $^{20}/_{100}$ of whatever distance the train actually travels is the **measurement aberration** for distance travelled by the train with respect to and due to the motion of the observer.

For the observer to physically perceive this discrepancy the distance travelled by the train must be measured with an accuracy much better than 20/100. Why? Well, let us consider what would happen if the accuracy worsened to say 21/100. In this case, when the train travels a distance of 100km we can only reliably say it has travelled 100±21km. Meanwhile, the distance travelled by the observer will be 20±4km. The cumulative error of the sum or subtraction of the two is ±25km. Now, obviously when a distance aberration of 20km is observed as above with an accuracy ±24km the observer should realize that they have wasted theirs and our time observing the aberration in the first place because the accuracy is no good compared to the magnitude of the quantity being measured. This situation makes about as much sense as measuring the length of a meter rule to an accuracy of ±1.2m (in effect implying that the length of the rule might actually be zero or even, as you might have guessed -0.2m (i.e. 1 - 1.2 = -0.2) long which is of little meaning considering we can physically see the rule we measured with our own two eyes.

Now consider the beam of light mentioned earlier travelling at 300000km\s = 1,080,000,000km\h. Our observer is still travelling at 20 km\h. The observer will perceive the beam of light to have travelled 20km less than the actual distance it has travelled for every 1,080,000,000km it does travel. In order to perceive this discrepancy we must be able to measure the distance travelled by the photon beam in an hour with an accuracy better than 20/1,080,000,000 as per the previous argument. This is equivalent o measuring a distance of 54,000,000 km to an accuracy of ±1km. For reference purposes this distance is equivalent to 180 laps around the earth's equator. There are currently no known means of measuring such a vast distance with such accuracy. In fact, when such distances are stretched out into space they are commonly quoted to an accuracy of ±1000km. Now again from the laws of error propagation the difference between the distance travelled by the observer and that travelled by the photon beam 20km±1000km. this demonstrates that there is no way, with the technology currently available to us, that such an accuracy can be sanely used to measure a distance of 20km. Therefore, unless we can find a way of improving the accuracy with which we measure such distances by a magnitude of 100 then there is no way the observer will be able to detect the discrepancy between the subjectively perceived speed of the beam of light and the actual speed of the same beam.

Ok... that's all well and good but why use such a great distance? Why not bring everything down to a distance we can measure to that accuracy. Surely that will sidestep the problem... right?

At a terrestrial level GPS is the most accurate means of measuring large distances. It can pinpoint a location on earth to an accuracy of a few meters so let us assume our beam of light runs for one lap around the earth's equator. Given a mean Geodesic representation of the shape of the earth which this beam follows we can optimistically assume that the distance travelled by the beam is known by GPS to an accuracy of ±1m. So the distance travelled by the beam of light is at best 40,080,295±1m.

Now, Light will travel this distance in 40080295/299792458s i.e. 0.1337 seconds. In that time the observer travelling at 20km\h would have travelled (20/(60*60))*0.1337km i.e. 0.743m. Now... at its very best the GPS receiver he holds as he follows the beam round the earth will know his position to ±1m. The distance travelled by the observer is therefore given by 0.743m±2m. Another nonsensical result.

So you say fine... maybe the problem is that the distances we have been using up till now are too big. At smaller distances we can measure distances to a much greater accuracy than 1m which would mean any movement on the part of the observer would be detectable. If we use a 100km distance light will cross this gap within 100/1,080,000,000 hrs (i.e. 9.25 ×10^{-8} h = 0.0003s). In that time our observer travelling at 20km\h will have covered a distance of 1.6mm. Attaining such an accuracy given these circumstances would be considered a monumental task. For the time being and most definitely in 1905 this level of accuracy for measuring the motion of the observer is not available. So for all perceivable/practical purposes this accuracy can not be achieved today and the difference between the speed of a photon train as perceived by the observer and the actual speed of the train will be undetectable and hence a big **zero**.

The important thing to note here is that we can not conclude from this that the speed of light is absolutely constant irrespective of motion. You can only deduce that the difference between the observed speed of light and the actual speed of light is imperceptible given the limitations of our perceptive abilities aided by modern day technology. Equating it to zero will be consistent with reality as we perceive it but **not** as it really is. Obviously since we don't know what reality really is, it is perception that matters for all practical purposes so any experiments done to verify special relativity will remain consistent with this zero aberration erroneously seeming as though they prove the absolute truth of this theory.

In fact we can never observe the change in the velocity of light due to the motion of the observer until that observer travels at a speed such that for a certain fixed (measurable) time interval the proportion of the distance he has travelled to the distance travelled by a beam of light is greater than the accuracy with which the distance travelled by the beam of light can be measured. Say the distance travelled by the photon beam can be measured to an accuracy of 1m by GPS for every 100 km (i.e. if the photon train travels 100 km we know this is true to the nearest m). The accuracy is $1/100000 = 0.00001$. Now the observer must travel a distance of about 4m in this time in order to avoid the nonsensical conclusions we came to above. The distance aberration will be 4±2m which can be perceived with the prevalent accuracy limits. The jockey must hence be travelling at $4/100000^{th}$ the speed of light or 12 km\s (i.e. 12000m\s). This is equivalent to Mach 23. By way of comparison even the Concord could only manage a maximum of Mach 2. Even if we could get something up to that speed we will only just be able to notice the aberration within a carefully controlled experiment.

We can conclude from all of this that the speed of light is not necessarily constant irrespective of the motion of the observer but that the discrepancy that arises due to the motion of the observer is so that it is imperceptible. In other words, for all practical purposes and within our perception of reality, the speed of light can be taken as constant without any cause for alarm. This is due to limitations imposed on our perception of the world around us using our five senses and not some universal law aligned with the special theory of relativity. However, the only way we can conclusively prove the theory wrong is to look beyond these limitations using much better and more accurate means of observing and measuring the world around us than we have today. This is sure to happen eventually, it's just a question of time.

6.2.3   Is the speed of light a physical speed limit?

Another assertion made by Einstein's Special Theory of Relativity is that the speed of light is a cosmic speed limit. This assertion logically follows from the notion that the speed of light is constant as discussed above.

If we assume that the speed of light in a vacuum is constant we are implying that light in a vacuum can not be accelerated or decelerated.

If it could; then the speed of light can not be held as universally constant in a vacuum.

Now if I want to accelerate a car I step on the gas peddle. Doing so indirectly causes all cars coming towards me to appear as if they too have accelerated by as much as mine towards me, even though their drivers might not have accelerated them. Special relativity says that not only is it impossible to accelerate or decelerate light but it is also impossible to perceive a different speed of light due to motion on the part of the observer. If this were not held true then theoretically, in a similar fashion, the speed of light relative to an observer can be indirectly accelerated **as** a consequence of motion on the part of the observer.

The logical argument for the universal speed limit continues from this as follows: If you can not accelerate or decelerate light you also can not accelerate anything traveling a bit slower than the speed of light all the way up to the speed of light. Counter intuitive, yes, but that's the theory. The reverse is true such that bodies traveling close to the speed of light can not be decelerated making the speed of light the ultimate speed limit. It would appear that if something did approach the speed of light it would literally be stuck there unable to decelerate back to normal or accelerate any further. If this does not happen then the Special Theory of Relativity will be contravened and we can't allow that can we?

To help nail this down what Einstein did was to observe that the faster an object appears to move the "*heavier*" it feels to a force trying to stop it. This is commonly known as inertia. The force, as it tries to stop the moving object, will expend a greater amount of energy, increasing with the speed of the object. Einstein made a leap of faith here (which has paid off quite well) by stating that as a moving object approached the speed of light its inertia approached infinity. Any force trying to stop this object will require an infinite amount of energy to do so and logically, the reverse argument that it takes an infinite amount of energy to get the object to this point must also hold. By this sleight of hand genius the theory of Special Relativity became asymptotic on approach to the speed of light and everything made sense, or did it?

6.2.3.1   Perception is reality

You and I perceive the world through five senses. Every now and then we aid and assist these senses to perceive more detail about our reality whether it is by a microscope, a telescope, a hearing aid or even by adding hot sauce to your food. The senses of Touch, Taste, Smell, Hearing and Sight are constantly feeding you with an unrelenting flood of information about reality. Just because these are the only conduits of information available to you, it does not mean that they are the only ones that exist in absolute reality.

Of all your five senses, sight is arguably in command of a disproportionately larger share of the information you gather from your surroundings than the other senses. This is both in terms of the quantity of information it gathers at a given instant and the information throughput it handles. If you consider the speed of light to be 300000km\s then every second you keep your eyes open you will get a 300000km long stream of information impinging on your retinas. This information is divided up into photons and you will be tasked with processing about 10^{18} photons in this one second because in the next second you will have another 300000km long data stream to consider. This ability of light to tell us a seemingly infinite amount about the world around us has been at the core of arguably all major scientific advances, from the telescope to the microscope to modern day laser technology prodding at individual atoms. Light and hence the sense of sight has become a lord of experimental science.

All this having been said; we are still limited in our ability to milk light of all the information it holds due to limitations in our sense of sight. Even when we aid it with microscopes and binoculars it seems there is still a lot more information we are not able to trap or process. Maybe in the future our eyes will develop in line with our brain, increased capacity to process more of this information. Unless Mother Nature grants us this vast, extra capacity we shall remain in a way, visually incapacitated.

It should not come as a surprise that many of our theories of reality are limited by our restricted perception of the world around us. Because light is the most detailed conduit of information we have, any limitations imposed on us towards obtaining information from it are implicitly limitations in our ability to understand the true nature of reality. It is from this foundation that you can begin to appreciate flaws not only in the theory of relativity but in all theories and models of reality put forward by man.

Those that fail to realize and accept these limitations end up fighting a loosing battle, trying to argue that a given theory or model (e.g. the big bang) is a true representation of reality. Those that realize and accept that these restrictions exist are in a better position to improve on mans current interpretation of the world around him.

6.2.3.2   The man on the Bus

Now imagine this. You are in a queue about to enter a bus and the man ahead of you is at the door about to get in. As he raises his right foot to step aboard you are suddenly endowed with the ability to travel faster than the speed of light. In fact let us blow reality clean out of the water by imagining that you start to run round the bus at 100000 times the speed of light. If you failed to imagine that scenario then pat yourself on the back because you are the smarter one amongst many. If you were successful in imagining that scenario let me disappoint you by letting you know that whatever you imagined in your head is far removed from what would actually happen. In fact, I too do not know what would happen. What in the world is he on about? You ask. All will become clear by the end of this section.

Let's start with something closer to home and build the case up to the scenario above. You are seated in front of the TV watching your favourite program when your partner calls your name. You decide to stand up and run to the pub but your partner keeps calling your name from behind. All of a sudden a Jet fighter swoops down, picks you up by the collar and carries you off. Now imagine you have some means of amplifying the calls from your partner still back there repetitively and unrelentingly calling your name while the Jet fighter screams towards the speed of sound.

The sonic boom thunders to herald the jet fighters passage into Mach 1 territory and you realize that you can no longer here your partner. There's nothing wrong with your super hearing and you know for sure that your partner will not stop calling your name until you get back but no matter how much you try you can no longer here the calls. You breathe a sigh of relief as the Jet fighter holds at Mach 1 for a few minutes.

The reason you can no longer hear the calls is because you are now traveling at the speed of the conduit of information (sound in this case) that tells you about the calls. So long as you continue traveling at this speed none of your partner's subsequent calls can reach you. In fact, you can no longer use your sense of sound to perceive the world around you as you could before. I'm not saying you can't hear anything but that whatever it is you hear will not represent reality as you used to know it. You are for all practical purposes like a deaf person now even though there is nothing wrong with your ears.

As if you thought things could not get any worse the jet fighter decides to scream on well beyond the speed of sound towards Mach 2. On the way something strange happens you start to hear you partners calls again but this time your name is being said backwards. At Mach 2 everything sounds the same but you name is still being said backwards. This is because you have caught up with sound waves released earlier by your partner and you are obviously hitting them from behind.

As the Jet fighter screams towards Mach 5 you here the calls more frequently but they remain reversed until you here the last call which coincidentally was the first call you heard while watching your TV after which everything goes silent. It's important to note here that your partner is still calling your name and will not stop until you return however you are now ahead of the very first call that was made and are traveling so fast that the calls can not catch up to you.

This should illustrate a very important point. As you travel up to and beyond a speed comparable to the speed of the conduit of information which you are depending on to sense the world. This conduit of information (in this case the sense of hearing) is rendered next to useless in its ability to help you continue to perceive reality as we know it. In the illustration above you can try to imagine how reality would be morphed however we have not considered how your senses would react and translate this information or even if at all they would continue to have information to react to and if so what reality that information would paint. So whatever we imagine will happen is only at best 90% of the story until we actually conduct the experiment.

This brings us back to the man on the bus. As you whiz round the bus at 100000 times the speed of light the man in front of you in the queue will physically appear to remain frozen in the position about to board the bus. Light too, will appear frozen in space to you (i.e. a photon will not move noticeably from its position before you come back to it for another lap). That begs the question, what will you perceive is reality and even more importantly, **how** will you perceive it? You can't use light, because it is not being given enough time as you move around to interact with anything let alone your eyes. You must use a conduit of information not only faster than light but faster than yourself otherwise you will find it difficult to move about without bumping into stuff. All in all Light has been rendered useless as a conduit of information about reality as we know it and the reality you perceive as a result will be far different from this, however you perceive it.

6.2.3.3   Traffic cops

Now it is possible to understand why Einstein so wanted the speed of light to be the ultimate cosmological speed limit. It was because he realized the world could not be perceived as we know it beyond that boundary. It therefore made sense to construct an impenetrable wall at that boundary using the theory of Special relativity if anything, to maintain some sort of order within science. By saying the world does not exist beyond that boundary we can avoid having to answer or solve seemingly ridiculous questions born of the fact that no one knows how to reach the boundary in the first place much less cross it. As a reward, Einstein gained praise and recognition for providing a simplified and accurate model of reality as we perceive it.

Ironically, Einstein's definition of reality through the Theory of Special relativity holds water because reality as we perceive it will change completely when you cross Einstein's boundary. However, there is no reason to conclude that there is nothing on the other side of the boundary and that the boundary can not be crossed.

There is an analogy to be drawn here with events that occurred in the 16th century. In 1548 an Italian Monk, Bruno Giordano, was burned at the stake for Heresy in Rome. His crime? Being an early supporter of Nicolaus Copernicus's theory in which the religiously held and promoted believe that all heavenly bodies revolved around the earth was deemed wrong. The crime and subsequent punishment seemed just; after all, when the Roman Catholics looked to the sky sure enough the sun always rose from the east and set in the south, revolving round the earth just like God had ordained. Only a man possessed by the devil could believe otherwise and sure enough, such a man deserved to be sent to hell. Then, reality was represented by a simple earth centric model that worked well and there was no need or even tolerance towards changing it.

In 1571 Johannes Kepler discovered the three laws of planetary motion that not only proved Copernicus right but were the basis on which Isaac Newton's universal laws of gravitation were founded. Mr. Giodarno's killing now appeared more of a crime than his heresy. The priests that sent him to the stake must have sat in their chambers dreading the day they would stand before God and attempt to justify their actions. The lesson to be gained from this is that just because a theory or model seems to represent reality well enough for current practical purposes does not mean it is the absolute truth. In fact, as discussed in the chapter on Models, by its very nature it can only represent but not be absolute reality.

The speed of light is therefore not so much an absolute cosmic speed limit as it is a traffic cop chasing after you warning you to slow down for your own safety because towards and beyond the boundary lies uncharted territory. There beyond lies a world for which you are not engineered to perceive or live in. Though no one has ever managed to go across we know of nothing that will stop you going across however if you do go across you are on your own and we have no guarantees you'll come back the same, or even that you'll come back at all.

6.2.4 Experimental proof

> "...It's a brave generalization, but
> it has implications that should in
> principle be observable."[21]

There are many experiments quoted by scientists as proof of Einstein's theory of special relativity. I will openly discuss some of them below. It is important to note that none of these directly prove the theory of special relativity to be correct. They are simply observations which are consistent with the theory and with all honesty I'd be surprised if an observation made within the limits of the speed of light will ever contradicts this theory. The theory models the world as we see it and the only means by which it can be disproved is to perceive the world as we currently **can not** see it by observing motion at and beyond the speed of light. Something that is not yet possible so no doubt Einstein will reign supreme for a while to come.

6.2.4.1 Atom bomb $E=mc^2$

The Atomic bomb that ended World War 2 was obvious proof that there is more energy within matter than meats the eye. How much energy exactly? No one knew, however, Einstein's 1905 equation $E=mc^2$ (which ironically was placed at the end of the paper almost as if to brush it aside as a less important afterthought) gave a figure high enough to contain the observed energy and

---

[21] João Magueijo in his book "Faster than the speed of light" p36 on the implications of Einstein's theory of Special relativity.

thereby avoid any obvious challenge. No one could measure the actual total energy released however all figures obtained were well within the predictions of Einstein's equation.

Now $E=mc^2$ follows logically from Einstein's argument that the inertia of a body tends to infinity as the body approaches the speed of light. Einstein then infers that imparting Kinetic energy (motion) onto a body increases its inertia (which is proportional to the bodies mass multiplied by the bodies speed). But at the speed of light his theory predicts that the body will have an infinite inertia as discussed earlier and since the speed of light is finite and known it must be the mass that becomes infinite and in so doing, generates an infinite inertia. So it therefore follows that at the speed of light a bodies mass must tend to infinity in order to keep Einstein's theory of special relativity happy. In other words the kinetic energy being imparted on the body to accelerate it up to the speed of light is being converted to the mass gained by the body. But why should kinetic energy be important why not say that any energy can be converted to mass. In fact, why not say that energy and mass are equivalent and that given a particular mass we can multiply it by a constant of proportionality and derive the amount of energy it contains. C^2 seems to give a realistically high enough value so why not write **$E=mc^2$**.In all fairness there is a bit more to this formula than this. Some mathematical derivation was involved which was later published by Einstein to reinforce the theory (see chapter 8.2.2). When one walks through Einstein's derivation of the equation one is struck by what appears to be a fatal flaw in which Einstein exploits the fact that any number multiplied by 1 remains unchanged, using it to introduce the speed of light into the equations of momentum[22] and hence, by sleight of hand (or pencil for that matter), obtains the famous $E=mc^2$.

So the atomic bomb was not a predicted consequence of Einstein's famous equation but more something that simply fits in with the equation just as much as $E=mc^3$ would have. Unfortunately we do not have the technology and means to independently measure all the energy contained within a unit mass. If we did, it should not come as a surprised if the figure turns out to be substantially higher than that predicted by $E=mc^2$.

6.2.4.2   The Michelson Morley Experiment

This is arguably the most famous failed experiment in history. The experiment was conducted in the 1880's in an effort to detect the motion of the earth through the "*ether*" by measuring the difference in the speed of light determined along the line of the earth's motion and at right angles to that line. In so doing the experiment would prove the existence of an ether wind which would imply that the speed of light is dependant on the motion of the observer relative to this ether.

---

[22] Mark McCutcheon explains this in great and eye opening detail in his book "**The final theory**" p333 to p336. Though there are a number of aspects in his book with which I do not agree he shows an uncanny talent for logically breaking down some of the myths surrounding science today.

Einstein himself said he was not aware of the Michelson-Morley experiment when he came up with the special theory of relativity attributing his inspiration to James Clerk Maxwell's equations on the motion of electromagnetic waves (in which the speed of light was held as an absolute constant of course). Scientists however still quote this experiment's failure as having proven the constancy of the speed of light. As discussed in chapter 6.2.2, all this experiment proved was that any aberration in the speed of light due to the motion on the part of the observer could not be detected due to limitations in perceptual/ measurement accuracy. Some scientists even argue that the results of the experiment were inconclusive and Michelson and Morley might yet have the last laugh on the matter. All in all, there is no doubt that Einstein must have been pleased at the experiment's failure which implicitly reinforced the foundation under his Theory of Special Relativity.

6.2.5   Conclusion

For the reasons above Einstein's theory of Special Relativity has a Jekyll and Hyde influence on scientific progress. On the one hand it lays down fundamental principles by which reality as we perceive it behaves. These have been used as a foundation for more elaborate models of reality leading to some useful applications. On the other hand it is inherently based on assumptions that seem to impose limitations on science and research which might not even exist (i.e. the constancy of the speed of light and the assertion that nothing can travel faster than this). These currently can not be disproved experimentally due to limitations imposed on our perception of reality as discussed above. Consequently we are left with a logically weak theory which though it is inconsistent with many aspects of Newtonian physics helps to model other aspects of reality not accounted for by Isaac Newton.

This conundrum is not new to science. In fact, the very first ever recorded scientist, Thales of Miletus (about 625 BC) was a philosopher who proposed that the world was made out of water and that the earth is a disc that floats on the water. There were no star maps, telescopes or ships sailing vast distances to provide a means for anyone to contest this model. Even though 11 years later other philosophers started to suggest that the earth was round it was not until about 580 BC that the likes of Pythagoras actually taught contrary to Miletus that the earth was a sphere. Ever since then, through the discovery that we are not the centre of the universe, and then that atoms are not exactly billiard balls and into today's Quantum weirdeties; Scientists have put forward model which though they did not describe absolute reality, described it sufficiently for the means and purposes of the time. Historically, this would only be uncovered by the generations that followed as they pushed the boundaries of human perception and experimental investigation. Einstein is no different from Thales of Miletus, Ptolemy, Galileo or Isaac Newton.

The disappointment here is that the institution of science has gradually been turning its back to the ways of the heretics of old that progressed it by leaps and bounds to the heights it scales today. The institution is choosing to head down a path strewn with religiously held and unchallengeable believes in individuals, theories and models. The same religious fervour that threatened to kill science back in the days of Ptolemy and Galileo threatens to kill science today through the pear review process.

6.3 The Mysteries of Light

The double slit experiment is one of the great mysteries of light. The experiment was made famous through a paper published in 1802 by Thomas Young on the wave theory of light. In this paper Mr Young described an experiment by which light could be made to produce an interference pattern much like the ripples in a pond.

From these experiments he logically concluded that light is better described as waves than the particles Isaac Newton suggested it was. The irony of the story is that Einstein's photoelectric experiment flipped the tables back in favour of Mr Newton and ever since then light has been accepted to have both particle and wave characteristics. This dual wave-particle nature of light and its mysteries will be discussed in detail in chapter 7.5.

7 Quantum Physics

Quantum physics is the umbrella under which scientist are continuing to try and understand the nature of the smallest entities we know of. As a division of science it opens out into a seemingly fantasy world littered with mystery and strange, unexplained happenings.

7.1 Humble Beginnings

It was in 1900 when Max Planck, through what seemed like guess work, quantized energy. The result was a simple now famous equation where the energy intrinsic to a 'piece' of radiation (**E**) was related to the frequency of this radiation (**v**) by a constant of proportionality (**h**) from then on known as the Planck constant.

$$E = h \times v \ldots\ldots\ldots\ldots 7\text{-}i$$

Note that this formula was not arrived at by measuring both the frequency and total absolute energy properties of a large set of pieces of radiation. It was essentially a lucky guess which seemed to fit in with reality as was perceived. In fact it fit in so well that it solved a riddle which up until then had danced around all the greatest minds of his time, the Ultraviolet catastrophe[23].

This equation solves the ultraviolet catastrophe by dictating that energy only exists in distinct units called quanta whose value is invariably given by equation 7-I above. If the frequency increases then the quanta of energy required to constitute that piece of radiation increases meaning that only a few of these pieces of Electromagnetic energy can be emitted. A lot of low energy frequencies are emitted however they have such a small amount of energy that as a whole they do not amount to much. However, the middle frequencies have enough energy and are numerous enough to cause the peak in what is now known as the blackbody curve (see Figure 7-1).

---

[23] The Ultraviolet catastrophe resulted from the classical assumption that electromagnetic waves behave like string waves. As such the laws of statistical mechanics dictate that the energy released at a given frequency is proportional to the frequency, implying that when a heated blackbody is observed there should be an infinite amount of energy emitted at the lower wavelengths (higher frequencies). This was contrary to observations that the amount of energy appeared to tail off as the frequency of the radiation produced increased.

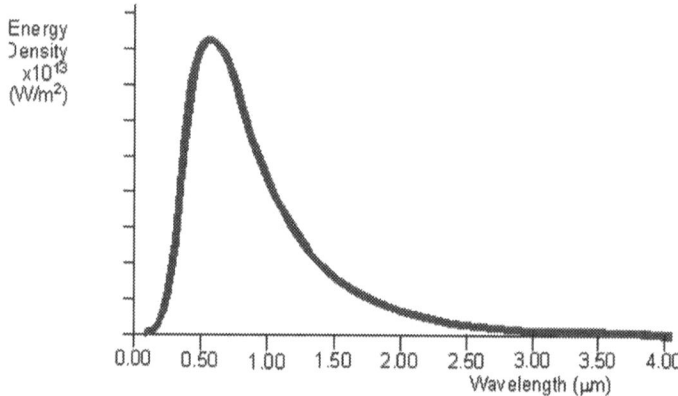

Figure 7-1: Theoretical blackbody curve

Yes, as before, the energy of the electromagnetic radiation was still proportional to the frequency of that radiation but by quantizing this radiation as per the explanation above Max Plank was able to explain how the blackbody curve tails off from a peak in the middle towards lower values for higher and lower frequencies. This quantization was the humble beginnings of the science of Quantum Physics. Today it has given birth to such diverse fields of study as Quantum Computation, Quantum Chromo Dynamics and Quantum Mechanics as more and more scientist try to expand on the foundation set by Professor Max Planck when the word "*Quanta*" was first used.

7.2 Does god play dice?

7.2.1 Atomic roots

In 1916, bathing in world fame and acclamation, Einstein himself introduced statistics into the study of the very small. As everyone was desperately trying to digest, understand and hopefully test his general theory of relativity, Her Doktor Einstein turned his attention to Bohr's model[24] of the atom. In particular Einstein was interested in the <u>probability</u> that a particular Bohr Atom in a given quantum state would deteriorate into another different known quantum state and in so doing emit some electromagnetic energy. By applying standard mathematical statistics he was able to derive equations representing the behaviour of a large number of such atoms. These were found to be identical to Planck's equation. In so doing Einstein seemed to prove that the Bohr model was consistent with other prevailing physical theorems and phenomena. He thereby fostered the acceptance of the model and Bohr's subsequent Nobel Prize. Little did he know that he would spend his dying days fighting to undo the consequences of this root foundation.

---

[24] Published in 1913 the Bohr model provides arguably the most familiar image of the atom with electrons 'in orbit' around a central nucleus much like the planets orbit round the sun.

7.2.2   Enter the Quantum age

Now, by 1916 Einstein was well acquainted with Max Plank's idea of Quanta having published a paper explaining the Photoelectric effect in 1904[25]. In fact for the 10 years since this paper, another scientist, Robert Millikan, struggled to disprove Einstein's quantization of light but ironically, in 1916, ended up admitting that his research seemed to prove this very phenomenon. In retrospect, it seems as though Einstein saw this as a prompt to further build on his now substantiated theory concerning the quantization of light. He began looking at the momentum of these quanta of light. This was yet another momentous occasion in the history of quantum physics because it was to be used in his dying years to argue the case against some of the weirder consequences of the quantum physics he was helping to establish.

After a bit of a lull, in part due to the First World War, scientific research came back with a vengeance in the early 1920's. In 1924 and 1925 there was a flurry of activity which rapidly built on the idea of the "*Quanta*". The period saw such great names as Louis De Broglie[26],

---

[25] The photoelectric effect had already been observed and established independently by Philip Leonard and JJ Thompson. Leonard's experiments in 1899 and subsequent investigations had proved the quantization of light however it was Einstein in 1904 who applied Planks equation directly to light hence formally starting the idea that light traveled in discreet units.

[26] Formally introduced wave-particle duality which lies at the heart of quantum physics.

Satyendera Bose[27], Wolfgang Pauli[28], Werner Karl Heisenberg[29] and Paul Dirac[30] place there mark on the development of Quantum theory. The term photon was first formally used to describe Einstein's particle of light by an American Physical Chemist, Gilbert Lewis in 1926. Its arrival appears to have formally heralded in the beginning of a new scientific age. The age of Quantum Physics.

7.2.3   Quantum physics gets weird

It was in 1927 that that the era of disharmony between Quantum physics and Einstein's work began. After discovering the central role uncertainty plays in the quantum world, Heisenberg nailed down the fundamental principle that would eventually bring Quantum physicists in a head to head clash against the authority of Her Doktor Einstein. It was this, Heisenberg's uncertainty principle[31] that

---

[27] Worked with Einstein to develop statistical rules outlining the behaviour of Bosons (particles similar to photons).

[28] Established Pauli's exclusion principle which states that no two fermions (e.g. electrons) can occupy the same quantum state. Electrons in an atom must occupy different energy levels and not congregate at the lowest energy level.

[29] Founded matrix mechanics which was the first complete and self consistent theory of Quantum Mechanics.

[30] He proposed his own version of quantum theory known as operator theory or quantum algebra.

[31] The principle states that we can never know both the instantaneous position and momentum of a quantum entity such as an electron. Knowing one more accurately introduces uncertainties (errors) in the other quantity hence rendering it's determination to any meaningful degree of accuracy impossible.

ushered in the spate of weirdeties we've come to associate with the quantum world.

The uncertainty principle does not only exist in the Quantum world, but is also evident in the world more familiar to you and me. For example, playing pool and armed with a calculator and a few basic mathematical formulae the average Joe can predict the position and momentum of the moving balls that will result from a particular, precisely placed shot. Heisenberg's uncertainty or error, which is of the order of $\pm 10^{-27}$ still applies but is irrelevant considering the size of the quantities being measured. If all we need is a distance accuracy of ± 0.001m to perform the required shot an error of $\pm 10^{-27}$m will have no consequence on the shot's outcome in any way perceivable. It is only when we get to very small measurements that the uncertainty error rears its ugly head. In fact it gets worse, if we try to use these small measurements in equations. For example, if ρ_e (momentum of an electron) is measured we find that this error becomes so significant it forces us to talk in terms of the <u>probability</u> that an electron is in a certain location. In a way we start to admit the truth that we do not truly know where the electron is. Being the stubborn beings that we are, we refuse to be beaten and employ statistical probability to counter this and give us a degree of confidence about the electrons location. This is pretty much the last kicks of a dying horse. Ironically, the statistics employed can be extrapolated to imply that the electron can be anywhere in the universe just that it is highly more likely to be in one place rather than another. We then feel justified in making statements such as "*there is no absolute truth at the quantum level*"[32]. This is where what started of as a precise science ends up in a world of exotic theories which conjure up an image of a fantasy world that is separate from reality as we know it. This sure makes for sensational science but is it progressive science?

7.2.4 Quantum physics becomes formalised

The three years up until 1930 saw a concerted effort to formalise Quantum theory. There were a number of interpretations. Paul Dirac proposed a version which took account of the requirements of Einstein's theory of special relativity while Heisenberg and Pauli proposed the langrarian version. It was however the Copenhagen interpretation which held sway into the 1930's and stood unchallenged up until the 1980's. This dominance was not achieved through it's being the '*true*' interpretation, far from it, no one even knew what a '*true*' interpretation of the weird quantum reality was. The Copenhagen interpretation bore a lot of flaws however it achieved it's dominance largely due to the forceful personality of Niels Bohr and was named so because Bohr worked in Copenhagen at the time.

The Copenhagen interpretation among other things brought attention to the inherent

---

[32] "In Search of Schrödinger's cat", John Gribbin, Black Swan edition 1991 p120.

limitations of probing the quantum world and translating any information gleamed from it into how we perceive and understand reality. It demonstrated that any such attempt would produce weird outcomes. In illustration of this and stressing the importance of experiment, Bohr explained that all we really know is what we measure with our instruments and the answers we get depend on the questions we ask. These questions are influenced by our everyday experiences. We design experiments which are a part of the world of classical physics which we know does not describe the quantum world and continue to use these to ask question within the quantum world. Because we can only know what is happening in the quantum world by probing it with our instruments, it is meaningless to say what quantum entities are doing when we are not looking at them. All we can do is calculate the probability that a particular experiment can come up with a particular result.

In his book "The Cosmic Code", Heinz Pagels (at the time president of the New York academy of Sciences) expressed Bohr's interpretation as follows:

"*...there is no meaning to the objective existence of an electron at some point in space..., independent of actual observation. The electron seems to spring into existence as a real object only when we observe it... Reality is in part created by the observer*".

The formalisation of quantum physics continued however and in 1930, Paul Dirac published "The

Principles of Quantum Mechanics", which was the first systematic treatment of quantum physics. In 1932 John Von Neumann published his fallacious proof that no '*hidden variables*'[33] theory can describe the quantum world. Though Grete Herman exposed Von Neumann's flaws at the time, her work was largely ignored because of the strength of Von Neumann's reputation.

It was not until 1935 at 56 years of age that Her Doktor Einstein thought enough was enough and decided to formally step in and help sort out the confusion. He teamed up with colleagues Boris Podolsky and Nathan Rosen to formally return fire at the now sprawling Quantum Physics camp. The three published a paper drawing attention to what seemed a paradox of the quantum world, now known as the EPR paradox. Einstein would die 20 years later hopelessly arguing his point that "*God does not play dice*". One wonders whether it ever crossed his genius mind that it was himself that had brought the dice to the table in the first place back in 1916.

---

[33] The 'hidden variables' theory is an interpretation of quantum theory which suggests that there is an underlying layer of reality below the quantum level. This layer contains additional information about the world in the form of hidden variables. The argument is that if we knew the values of these hidden variables we would be able to predict the precise outcome of a quantum experiment rather than just the probability of getting particular outcomes.

7.3 The EPR paradox

7.3.1 So it Begins

In 1935 Albert Einstein, Boris Podolsky and Nathan Rosen published a paper[34] in response to growing exotic implications of the Quantum Theory that Einstein himself had helped bring about. This paper contained a thought experiment which from then on was known by the initials of the three progenitors as the EPR Paradox. The thought experiment aimed at challenging the contemporary thinking around Quantum Mechanics fostered by Heisenberg's uncertainty principle in which it was stated that it is impossible to know both the momentum and position of a particle at the same time.

Einstein and friends argued that if two quantum entities were to interact it would be possible to measure the momentum of the whole. These entities are then left to go their separate ways without interacting with anything else on the way. It is then theoretically possible to measure the momentum of the first particle and by subtracting this from that originally measured for the whole we implicitly know the momentum of the second particle. One could then measure the position of the second particle. In so doing Heisenberg's uncertainty principle would be violated since both the momentum and position of a quantum

[34] A. Einstein, B. Podolsky, and N. Rosen, "**Can quantum-mechanical description of physical reality be considered complete?**" Physical Review, volume 47, pp. 777-780, 1935.

entity at a given instant would be known. In fact, the only way for Heisenberg's principle to continue to hold would be if there was some way in which the two particles could communicate instantaneously over the separating distance such that the second particle knows when the first has it's momentum measured so as to '*hide*' it's position. This "*spooky action at a distance*" as Einstein, Podolsky and Rosen referred to it, seemed illogical.

The basic idea aiming to test this hypothesis was developed by David Bohm in the 1950's, refined by John Bell in the 1960's leading to an actual experiment in the 1980's that seemed to prove that this action at a distance actually occurred.

7.3.2   Einstein's logic vs. Heisenberg's principle

Werner Karl Heisenberg formally presented what is now known as the *Heisenberg Uncertainty Principle* in the form of a paper published in 1927. At the end of the paper he stated:

> "*We cannot know, as a matter of principle, the present in all its details.*"

According to Heisenberg we perceive the world around us by bouncing photons off the objects we see. On the macroscopic level these photons do not affect the object being observed e.g. a football. They are simply too small to cause any macroscopic change in the state or location of the football. They might cause some imperceptible microscopic changes e.g. miniscule changes in temporal temperature at the point of incidence, but these are not perceived by us and are irrelevant to the objects utility in the macroscopic world. All we really care about is that the football will stay in its state of motion or rest until it is played in one way or another. However, on the subatomic scale, objects are affected by photons much like what happens within a billiard ball collision such that attributes of the observed object (i.e. position) can be changed by the incident photons.

Say we collide a photon with a subatomic particle e.g. an electron in an attempt to know the position of the electron. The collision will interfere with and help change the electrons momentum (mass x velocity) rendering any information we previously had on its momentum useless in describing its state after the collision. It seems as though the act of observation has changed the observed entity.

In actual fact the electron is constantly interacting with photons (irrespective of whether or not it is being observed) so there is no way of knowing exactly where the electron will be at a given time. We can only state the electrons position to a given degree of certainty and say that it is highly probable to be in a certain region and highly improbable to be in other regions. This is the essence of the uncertainty principle.

The statement commonly made that the observer affects the observed is ever so slightly misleading because it is not the observer as such but the conduit of information (light in this case) which interacts with the observed object. In other words, there does not have to be an observer. At the quantum level objects are constantly interacting with the conduit of information (light) even when that light does not end up being observed.

What Heisenberg's uncertainty principle so vividly illustrates is that the world around you is defined in so far as you can observe it without disturbing it. Once the means you use to observe the world start to interfere with it you are rendered blind for all practical purposes.

To illustrate the point lets stretch the imagination and picture ourselves as gigantic beings, so big that the Milky Way galaxy is the size of an electron in the atoms that compose us. Now, as such beings lets imagine that traveling galaxies are the means with which we use to perceive reality and the world around us. At this point it is not important to understand how a perception of reality is realized but that the conduit of information in this case is the galaxy in much the same way that light was before. You and I can not imagine what reality such beings will perceive but we can imagine what happens when the conduit of information tries to report back to the being the presence of our Sun. The galaxy will interact with or even consume the sun such that after the observation is made any information previously known about the Sun is rendered useless in determining its current state. For all these beings know, the Sun could now be anywhere because the conduit of information (the galaxy) interacted with the entity being observed (the Sun). The only way this being can reliably observe our Sun is to devise a means of using a conduit of information that is much smaller than the Sun and does not affect the properties of the Sun that the being requires (e.g. position and momentum). We on planet earth already naturally use such a conduit of information, light. It limits our perception of reality in a similar way when we try to use it to observe the very small, however, it would be just what this being requires. It seems only logical to deduce that there is a sub quantum level conduit of information to which we as macroscopic beings are largely transparent and as such we do not have any knowledge of its existence much less any way of being able to perceive reality through it. The inability to directly observe the world through this sub-quantum conduit of information is what makes the quantum world seem weird.

If the huge beings mentioned earlier trying to observe our Sun realized that the Sun simply disappeared some of the times it was observed, they would conclude like we do about the quantum world that reality behaves different at this level from the way it is normally perceived to behave. However, at our level we know that the black hole at the centre of the galaxy they use as the conduit of information can actually swallow the sun up so the Sun does not simply disappear but becomes a part of the galaxy they use to observe it. It is important to note here that the quantum world is weird just because that is the way we observe it and not because that is the way it is. In other words, the world is

The Gods Of Science ■ 117

the way it is because that's the way we perceive it, if we were to perceive it differently, reality as we know it would most certainly be quite different and what seemed weird before would become a fact of normality.

7.3.3 Beyond the Limits of Human Perception

Because we can not perceive the quantum world through a conduit of information that does not affect the observed quantum entities what scientist have been doing is using the available conduits of information to bombard the quantum world. By statistically analysing resulting changes that can be interpreted on a macroscopic level they are able to get some macroscopic translation of the happenings at the quantum level.

Let's take the position of an electron as an example. We know we can not observe the position of an individual electron without affecting it but if we direct our conduit of information towards the general area in which we expect to find an electron, send out a multitude of information sensors (photons) towards the region then detect when a sensor has interacted with an electron (i.e. we observe a change in spin of the photon on detection) we can work out the probability of an electron being at a given location within the target region. But is this really what we are measuring. Could the conduit of information be influencing the electron in such a way that it distorts the distribution picture we get from reality. In addition, by resorting to probability we are implicitly admitting that we do not know where the electron is. The next scientist to pick up the problem from where we left off would be justified in saying that there exists an infinitesimal chance that the electron is on the other side of the universe and however small it is; it is a chance all the same. In addition, for two points **A** and **B** within the region being bombarded by a photon if an electron is detected it could be at **A** or **B**. In fact it is logical to conclude that the electron is at both **A** and **B** simultaneously since for all practical purposes (and assuming the photon does not affect the electron) any experiment we carry out again in this region will report the existence of an electron and not whether it is at **A** or **B**. This is how we arrive at the argument behind superposition of states[35]. In other words if we do not know what state the experiment is in then we assume that it is in all states possible, simultaneously.

A photon is polarized but we do not know the direction of polarization until we observe the photon. We can hence state with impunity that the photon is in all the possible, suspect directions of polarization just like we were able to state that the ball existed at both point **A**

---

[35] Superposition of states is when a quantum entity is said to simultaneously exist in two or more states (i.e. it has a quantum property e.g. spin, which simultaneously has two or more values e.g. both up and down spin). The superposed states will dissolve into one state when the property is measured and become superposed again afterwards.

and **B**. We have hence made a deduction from our lack of knowledge about the system that <u>only holds **because** of our lack of knowledge</u> of the system. If we could somehow observe the tennis ball or the Photon without affecting it we might find that at any given time it only exists in one state or position. Also, since it is constantly being influenced by the conduit of information, it is always changing this state or position. We can't observe it without affecting it so the deduction stands unchallenged <u>not because it is correct but because it can not be proven wrong</u>. This is like a religious preacher asking you to prove God does not exist and until you do it will be held as truth that he does. In fact, the only way to prove that this picture is wrong is by finding a means of perceiving the photon without affecting it. Something that is currently not possible.

Unfortunately, a whole science then grows around this misleading deduction taking it as gospel truth. In fact, it is observed that the deduction agrees with many pre-existing theories on reality including Einstein's theory of relativity. In fact why wouldn't it, after all it is based on reality as we perceive it just like all these theories. Be warned not to be tempted into the trap of trying to prove it wrong, like Einstein found out the hard way right up until his death bed. The more you struggle the more trapped you will become and the more you will realise the futility of your efforts. It's air tight because it does not violate the world and reality as we perceive it. This non violation is not because it is the true model of reality but because as a model it is not pertinent to the reality beyond our perception which in being beyond our perception can not be directly queried by any known experiment to an accuracy that will test this model.

7.3.4   Were Einstein's EPR efforts futile

So why can't we know both the position and momentum of a quantum entity. To explain this lets go back to the macroscopic exaggeration of the tennis ball. Imagine the tennis ball is moving undisturbed with a given momentum ($m_{ti}v_{ti}$). You want to determine its instantaneous momentum ρ_{ti} and position X_i. You decide to do this by throwing another tennis ball at the first and recording the time of impact T_i . By measuring the momentum of the target ball afterwards ($m_{ta}v_{ta}$) as well as the momentum of the object ball before ($m_{ob}v_{ob}$) and after ($m_{oa}v_{oa}$) the collision we can use the classical law of conservation of momentum to determine the momentum of the target at impact

$$\rho_{ti} = m_{ti}v_{ti} = m_{ta}v_{ta} - (m_{oa}v_{oa} - m_{ob}v_{ob})$$
$$\text{................ 7-ii}$$

Our ability to determine the momentum of the target ball will be influenced by how well we can measure the momentum of the object ball before and after the collision as well as that of the target ball afterwards (substituting these measurements in the right hand side of the equation gives us the required momentum). As with all measurements there will be a degree of accuracy associated with the measurement of momentum using the best instruments available to us which can be represented by $\Delta\rho$. Now, after the collision when these measurements are substituted into the equation the momentum of the target particle at incidence can be evaluated with an error of $\pm3\Delta\rho$. This is because according to the laws of propagation of error, when quantities are added or subtracted the errors within those quantities accumulate. It is important that the momentum of the target at the point of incidence derived from these measurements is of a much greater magnitude than this error because it will be written as $\rho_{ti}\pm3\Delta\rho$. In this case if ρ_{ti} is not much greater than $3\Delta\rho$ we might end up with a strange statement indicating that the momentum could be 0 or even negative which would not make sense seeing as we observe the ball is moving in the direction of positive momentum. It is therefore imperative that ρ_{ti} is much greater than $3\Delta\rho$ or else why bother with the experiment in the first place.

Therefore, we can write equation 7-ii as follows:

$$\mathbf{\rho ti = \rho ta - \Delta\rho o \gg 3\Delta\rho} \ldots\ldots\ldots\ldots \mathbf{7\text{-}iii}$$

Where $\Delta\rho o$ is the net change in momentum of the object ball (i.e. momentum after collision minus momentum before collision).

Equation 7-iii implies that the momentum of the target afterwards must be greater than the change in momentum of the object ball by an amount much greater than 3 times the error in measurement of momentum encumbering our measuring instruments.

Now let us consider two extremes of this scenario. Imagine that instead, it is the planet earth that is thrown at the moving tennis ball in order to determine the ball's instantaneous momentum. Because it is much bigger and has more inertia the change in momentum of the earth (ΔP_o) due to the impact will be very small in comparison to that of the ball. So the magnitude of the term P_{ta} - ΔP_o will be governed by the momentum of the ball after impact (P_{ta}) and will hence be relatively large. So large in fact that it will render the effect of $3\Delta P$ insignificant and in so doing enabling us to determine the instantaneous momentum of the tennis ball to a high degree of accuracy. In fact it is a general rule that the more inertia inherent in the incident object (i.e. the more reluctant it is to experience change due to the collision) then the more accurately we can determine the momentum of the target at the point of collision. However, more inertia implies more mass which (for all practical purposes) generally implies a larger spatial volume of matter. Though we can now determine the momentum more accurately the spatial extent of the incident object (the earth in this case) is too big to accurately determine the location of the target at the point of incidence. In fact, assuming the earth was thrown with Africa facing the direction of motion then the tennis ball could well have been hit by an anthill in Soweto or by boat under London Bridge. All we know is that the tennis ball was hit but it is pointless to try to locate the tennis ball since it could be anywhere within a circle the radius of the earth. This inaccuracy is so many orders of magnitude greater than the size of the tennis ball that the only sensible conclusion we can make from the observation is that the ball can be anywhere in this circle. In fact, according to the theory of superposition of states, before the collision the ball is anywhere and everywhere. It is only at the point of collision that the ball's location is determined and even then, the measurements taken to determine it (i.e. sounds from the area of incidence) can only give the statistical probability that it was in any given location within that circle. By measuring the momentum ever more accurately we have lost the ability to determine the location of the target as is stated by Heisenberg's uncertainty principle.

In order to better determine the position of the tennis ball we need to reduce the spatial extent of the incident object. We can do this by compressing it but short of picking up a rock from a super dense neutron star and escaping the gravity to tell about it we eventually hit a limit which dictates that we can only further reduce the spatial extent by removing matter (mass) from the incident object. This removal of matter means that a point will be reached when the incident object is so small that the change in momentum it experiences as a result of the collision is now of the same order of magnitude as the momentum of the tennis ball after the collision. In this case the term P_{ta} -

ΔP_o is no longer much greater than 3 times the accuracy ($3\Delta P$) with which our instruments can measure momentum and so errors in the measurement of momentum start to seriously affect the determined momentum of the tennis ball at the point of incidence (ρ_i). We can now determine the position more accurately but we have lost accuracy in the ability to determine the momentum of the tennis ball, again in consistence with Heisenberg's uncertainty principle.

7.3.5   What is really happening at the quantum level

In the world we are able to perceive there is a huge variation in density. In fact, by encapsulating air in a balloon we create an integral object of vast spatial extent but minimal mass. Throwing a marble at this balloon will allow us to locate its position and because the marble's inertia is large in comparison to the balloon it will not experience much of a change in momentum rendering the term P_{ta} - ΔP_o much larger than the error inherent in measuring momentum so we can also determine the momentum of the balloon to a comfortable degree of accuracy.

The Quantum world constitutes of the smallest entities we know which in turn are what macroscopic objects are composed of. These entities are so small that they do not interact with macroscopic objects in a classical manner. For example, if an electron is fired at a leaf of metal foil it will penetrate a few layers of atoms before it finally interacts with a particular atom if at all it does. In fact chances are it will pass right through the foil without interacting[36] unlike a tennis ball which would either bounce off the surface of the foil or punch a hole through it as per classical physics. In order to observe these classical momentum recoil behaviours we have to get quantum entities to collide with each other and not with macroscopic entities. This is where it all turns weird.

All the laws and equations we have inherited are consistent with and based on classical laws which describe the reality of the world we perceive around us. Everything from Kepler and Newton to Einstein's theory of relativity is based on observations made within the reality we perceive and of which we are part. We can not yet observe subatomic collisions, this has not prevented us from arrogantly extrapolating these classical laws into the Subatomic realm. We've created force carriers out of photons and Gluons[37]. Consequently, our original equations start to yield unexplained infinities and our experiments begin to yield weird results. This situation is telling us that either something is wrong with the assumptions we are carrying

---

[36] Rutherford, Hans Geiger and Ernest Marsden discovered this phenomenon in 1909 by firing alpha particles at a thin sheet of metal foil. Most of the particles went straight through but occasionally one bounced back from the foil. This observation was the basis of the Rutherford model of the atom.

[37] The quantum particles said to hold the atomic nucleus together.

over from the macroscopic to the quantum world or that the quantum world behaves different and weirdly in comparison to the world we can perceive.

It is much easier to assume that all assumptions that hold true about the macroscopic world should hold true about the quantum world which we can not directly perceive. This way we a comforted in believing that the reality we perceive is the **absolute** reality which accounts for even that which we can not perceive. The only way to challenge this logical deduction is to actually devise a means of perceiving the quantum world without interfering with it. This is similar to the argument that the only way to prove there is no God is to look beyond the end of the universe and then beyond the end of that which exists thereafter and confirm that you do not observe God. It is an argument that has the potential to go on for eternity without resolve.

The more difficult alternative would be to conclude that our classical laws and assumptions do not apply at the quantum level. This actually asks us to ignore our familiarities with classical world and devise new theories like Quantum Chromo Dynamics (QCD), Quantum electrodynamics (QED) and Quantum Mechanics which exclusively describe what we believe is happening within the quantum realm. In doing this we hope that as these theories mature, any differences with classical theory will be reconciled hence growing towards the holy grail of science. The theory of everything. Unfortunately, all these new theories are substantiated, and in fact can only be substantiated by classical physical experiments which interrogate the quantum world. As such the only way to mature them to a point at which we can say we understand the quantum world is for these experiments to be able to observe the quantum world without affecting it. Until then we are stuck with these abstract visualisations of the Quantum realm.

All of this would be fine were it not for those individuals who insist on prematurely and wildly extrapolating on these already weak theories to describe the Quantum realm in more detail. Before talking about teleportation, non-locality, quantum duality and the like it is important to appreciate that the quantum models from which these terms are derived are only that, models of reality. In fact they are very weak models of reality when compared to other models such as Newton's laws of motion because they are not based on direct classical (non-interfering) observation of the modelled reality. As such they should be extrapolated with even greater care and restraint if at all. If Newton's models which were based on actual non interfering experimental observation of reality, failed when they were extrapolated into space prompting Einstein to replace them with the curvature of space-time, why then should we expect that these new models based on classical assumptions and statistical observations of a world we can not perceive can be extrapolated with impunity and that the weird outcomes derived should be held as gospel truth?

7.3.6 Copenhagen vs. Hidden Variables

In the 1960's John Bell formulated an equation which could theoretically be tested through experiment to prove the existence of 'weird' quantum behaviour. This opened the doors for the Aspect experiment (see chapter 7.3.8). This equation and other subsequent equations are collectively commonly referred to as "Bells Inequality" equations. The inequalities are formulated such that if the observations made of quantum realm behaviour satisfy them, then the phenomenon of non-locality does not exist in the quantum realm. In other words, violating these equations verifies the principle of non-locality[38] and hence the weirdness of the quantum realm. In deriving the inequality Bell made the following common sense assumptions which if violated mean that the world is behaving in a way contrary to common sense:

1. The quantum entity attributes have definite values independent of the act of observation which account for all its observed behaviour. These well defined properties are known as hidden variables.

---

[38] This term is used to describe the way in which the behaviour of a quantum entity such as an electron is instantaneously affected by events at the location of the entity and by events elsewhere and everywhere in the universe. In so doing it contradicts Einstein's special theory of relativity which states that nothing can travel faster than the speed of light. Einstein spent his last days arguing against non-locality.

2. Physical effects have a finite speed of propagation in line with Einstein's special theory of relativity. In other words, physical objects can not exchange information faster than the speed of light.

Now, if the first common sense assumption holds then the superposition theorem is wrong. However, considering that the conduit of information (light) is constantly interfering with the quantum world, it logical follows that these quantum attributes are constantly changing due to the influence of this conduit of information (light) irrespective of whether or not there is an observer. In effect, statistically at any point, the attributes can be any one of the possible values as a result of these random interactions. This allows us to pull that old chestnut out of the bag saying that if we don't know what value it is then it must be every value possible at the same time. In effect, if this is so then the superposition theorem even though it does not explain reality (i.e. in reality every value does not exist simultaneously) holds firm and is not violated. It follows that the world we observe constitutes of instantaneous impressions of this dynamic reality.

If the second common sense assumption holds then Einstein's theory of relativity is not violated.

In general an explanation of quantum behaviour based on the Copenhagen interpretation assumes that both of these assumptions do not hold true and consequently have been violated (even though only one needs to fail in order to violate Bell's inequality). According to the explanation, when one of an entangled pair of superposed photons is detected, it instantaneously looses its state of superposition and communicates this instantaneously to its entanglement partner which does the same such that the two remain correlated.

Take note that these are common sense assumptions that were designed to test the then leading interpretation of quantum physics, the Copenhagen interpretation. If you remember this is not the leading interpretation due to merit but more due to the fact that it was the influential Bohr that proposed it. In other words, there is nothing to stop quantum physical behaviour being explained by a theorem in which both these assumptions hold and are not violated. Theoretically, a hidden variables theorem has the potential to do this.

Ironically, in some flavours of the hidden variables theory it is irrelevant whether or not any of these assumptions are correct. In other words, it is theoretically possible for a hidden variables theory to describe the quantum world but not violate Bells inequality. Are those gasps of shock at such a blasphemous statement that I hear? To illustrate this let us again imagine those giant beings to whom, proportionately, a galaxy is the size of an atom. One of the goliaths decides they want to investigate the constitution of galaxies (their equivalent to our atoms) so the being places a giant knife-like apparatus in the sky and somehow, through their experimental instrumentation, manages to hurl a galaxy at the blade of this knife. The being places detectors behind the knife that somehow register a rough approximation to the mass of the galactic portion that hits them. This is not observed as the mass we know of at our level of existence, but as some abstract quantum-like property X in their reality.

After throwing a large amount of these galaxies at the knife the being analyses the results to find that there is a significant amount of correlation between the readings from the detector behind the knife to the left and the readings from the detector behind the knife to the right. In fact, the giant being finds that statistically a higher value of X measured on the right will result in a lower value of X measured on the left at the same instance of time and vice versa. This being might conclude like our quantum scientists have that there is a superposition of states involved here and an instantaneous communication between the two sensors (non-locality) which facilitates the observed correlation of X. X represents a property of the galaxy which though the being can not measure it completely, will indirectly cause the correlation observed. In fact if the being could measure this property (i.e. mass) it would be possible to set up logical laws in the being's reality that predict the results of the experiment.

Now, we know, at our level, that the galaxy this being is hurling towards the knife like apparatus has a given fixed quantity of matter. If it is split in two then the bigger the quantity of matter in one part, the smaller the quantity in the other. There is no communication between the being's sensors or the two masses. The phenomenon being observed is simply a result of splitting an amalgamation of matter into two divisions which together must equal to the whole. If the being could directly perceive matter through light as the conduit of information just like we do then this would become obvious. But they can't so they too might be tempted to consider an explanation based on non-locality and superposition of states.

Though this is a rather simplified scenario it exemplifies how hidden variables theorems need not violate bells inequality in order to describe the quantum world. In this case there are hidden variables that can not be completely determined by the giant being and there is no super fast signal fostering instantaneous communication between the two sensors. Therefore Bells inequality does not need to be violated even though what seems like a weird correlation exists. The hidden variables interpretation is much more like common sense than the Copenhagen interpretation and more scientists are starting to take it seriously.

Let's dwell some more on the Copenhagen interpretation by discussing Bell's assumptions using a more classical, down to earth example. Suppose two observers, Alice and Bob, are each sent an entangled photon to observe. Bells assumptions imply limits on the correlation of the observations made by Alice and Bob on their photons respectively, in other words, we should see little if any at all dependency between the observation made by one and that made by the other. Any minute dependency observed can be attributed to freak chance.

If these limits of correlation are breached (i.e. there is a high dependency between the observations made by the two independently) then one or both of Bell's assumptions do not hold. In other words either information is travelling faster than the speed of light between Alice's and Bobs photon causing them to know about each other and hence behave in a correlated way or the attributes of photons do not have definite values (i.e. the values are ambiguous) until they are observed and at that point are simultaneously and instantaneously defined at both Alice's photon and Bob's photon irrespective of which photon was observed.

Bell's inequalities will no longer hold with correlated observations in which case they are said to be violated. It is this violation of Bell's inequalities by the Aspect experiments that lead to the conclusion that at least one of the assumptions made above must be wrong and therefore the quantum world behaves in a way contrary to common sense.

We've already shown that the conclusion of Einstein's theory of relativity stating that nothing can travel faster than the speed of light, only defines the reality we are able to perceive but has no grounds for argument in the reality that lies beyond our realm of perception. As such it can not be used to argue against the assumption that at the sub-quantum level quantum entities can exchange information faster than the speed of light.

The other assumption is slightly more interesting. The direct interpretation is that unless an observation is made by an observer the observed entity does not exist in any deterministic state. A slightly different interpretation appears to make things a bit clearer. The conduit of information we use to observe the world around us is everywhere all the time albeit in different concentrations. It is constantly moving and interacting with the real world whether or not there is an observer to assimilate the information it is conveying. In its interactions with the real world quantum entities it is constantly affecting them such that there characteristics, attributes and behaviour owe much to this conduit of information. In effect, without it the quantum world would be very different indeed. It is impossible to predict what this lightless world would be like without understanding the complexity of the infinite interactions that light has with other quantum entities at all levels. In fact, who's to say that a macroscopic change will not cause repercussions in the way light is interacting with the sub-quantum constituents of the macroscopic entity which would not only affect the local quantum world but would also instantaneously somehow affect the entire universe to an extent that decreases with distance from the local point.

A good analogy is that of air and sound. Sound moves through air at 330 m\s which is roughly the average speed of an air molecule at room temperature. Air as a whole however is normally not perceived to move much. Even so the speed of sound remains at 330m\s. In effect molecules of air can communicate to each other at a speed faster than there collective motion.

7.3.7   Is Statistics a real solution

In using a relatively vague statistical distribution to describe the location of the electron we have opened the door for even more vague interpretations of the electron's position.

To illustrate the point imagine an electron is the size of a tennis ball and the photon used to observe it is the size of the moon. The eager scientist then somehow manages to throw the moon at a region in which it is thought that the tennis ball resides not knowing whether the tennis ball is actually there or not. Now, the scientist can only prove the existence of the tennis ball by detecting changes in measurements made on the moon which indicate that it might have interacted with the ball. The gravitation of the moon might change due to matter dislodged from the surface during a really violent collision or it might be the general motion of the moon that is affected. The scientist makes these measurements but in the end, even if each individual moon is carefully observed as it is thrown at the tennis ball the scientist will only be able to assert that the ball does or does not exist within a region of space the size of the moon. The reality is that within that region the ball could be anywhere, however according to the scientist it either exists in that region or not.

There is no information available on where in that region it is at a given time so the scientist can only conclude that it is everywhere in that region at the instance of collision. Again as described earlier, reality has not changed but rather it is our interpretation of reality given the information available that has hit a limit and consequently caused us to misinterpret reality when looking beyond this limit.

If we throw enough moons at the region we can build up a statistical picture of the electrons distribution but we still will not know where it is.

7.3.8   Has strange Quantum behaviour been proved

The proof of the pudding is in the eating and the apparent proof of non-locality came through the Aspect experiments carried out in Paris in the early 1980's resulting in a paper published in 1982 by Allain Aspect and colleagues. These apparently established that what Einstein, Podolsky and Rosen called "spooky action at a distance" really does occur and that non-locality rules the quantum world.

An atom can be stimulated to emit two photons simultaneously and in order to remain consistent with the classical laws of conservation of energy the two photons must be of opposite polarization. Now, according to the Copenhagen interpretation, as each photon goes off in its own path we do not know it's polarization until we measure it. Again, the argument is made that the photon can be considered to be in both states of polarization otherwise known as a superposition of states and only when the measurement is made does it collapse into one state or the other. Of course we can then argue that the companion electron which could now be on the other side of the universe being in the same state of superposition somehow is notified that the first electron has been measured and automatically collapses into the opposite polarity without any measurement being made on it. This instantaneous response of the second photon to an action performed on the first was the action at a distance that the EPR paper argued does not make sense.

In the Aspect experiment two beams of entangled[39] photons were sent to the separate detectors from the source as shown in Figure 7-2.

A a   b B
∩   Photons   Photons   ∩
Detector Polarizer   Source   Polarizer Detector

**Figure 7-2: Schematic of the
Aspect experiment**

In their path was a set of polarizing filters whose angle was changed randomly throughout the experiment. The experiment was more complicated than the schematic above with 2 detectors on each side to detect both the

---

[39] Entanglement is the term used to describe a phenomenon in which two quantum entities are related such that if a measurement is made on one of the entities we can reliably predict the results of a similar measurement on the other entity without actually carrying out the measurement.

photons that went through the filter and those that were reflected by the filter. When the data was collected from each detector it was found that there was a significant correlation between the instantaneous polarization measurements at detector A and those at detector B. In other words, whenever a certain polarization measurement was detected at A the corresponding measurement at B could be reliably predicted with a high degree of confidence. High enough to discount the correlation as having happened by chance.

What was happening here is referred to as non-locality in which an event at point A seems to influence events at point B instantaneously. If the two points were communicating faster than the speed of light then we would be contradicting Einstein's special theory of relativity and that wouldn't be right. After all it is an absolute truth is it not? To Einstein's shock the Copenhagen interpretation explained this by saying that some communication occurs faster than the speed of light between the two points such that when one is measured and observed the other knows about this and takes up a known and dependant polarization state. Before this measurement both photons are thought to be in a superposed state. In the late 1920's this situation was tantamount to Quantum Physicist throwing tomatoes at Einstein for having lead them astray with his nonsensical theories for 25 years. No wonder Einstein came out fighting to his last breath intent on saving his legacy.

According to the hidden variables interpretation there is no superposition of states and if we knew the deterministic parameters of the quantum world we would be able to predict the outcome of a quantum experiment rather than rely on probability. In essence this requires no action at a distance. By knowing all the deterministic sub-quantum properties of the photon at the source in the Aspect experiment we can predict the measurements that will occur at the detectors. This is widely held as the more commonsensical approach to the problem. There is no mathematical or logical reason to discard the hidden variables theory. In fact John Bell proved that von Neumann's mathematical proof that no hidden variables theory could explain quantum behaviour was logically correct but yet the whole argument was based on an unfounded assumption[40], which no doubt invalidates the proof. The proof usually quoted against hidden variable theory can be simplified as follows: consider that equation 7-iv below represents a simplification of the Bell inequality.

$$y + y \neq 2 \quad \text{............... 7-iv}$$

This inequality is based on common sense assumptions that hold true for the reality we can perceive. If an experimental observation is made of the value y which violates this inequality then this observation is <u>not</u> consistent with common sense. Such observations are said to be of the

---

[40] "Q is for Quantum" John Gribbin, Phoenix Press, 2002. p214

phenomena of non-locality and if made, prove that the quantum world is weird.

Now imagine a hypothetical and similar inequality illustrating quantum weirdness as caused by hidden variables.

$$z+z \neq 4 \ldots\ldots\ldots\ldots 7\text{-v}$$

As previously discussed, hidden variables theorems need not violate the assumptions that lead up to Bell's inequality. In fact, hidden variables theorems <u>do not</u> appear to violate common sense (only that we can not observe them because they are hidden just like the giants could not observe galactic mass in the earlier illustration). So, based on this, if we place hidden variables in Bells inequality it is only logical that they will not violate the inequality (or common sense for that matter). For example, though $z=1$ will violate equation 7-v above hence proving quantum weirdness from the hidden variables perspective however substituting $y=1$ will not violate equation 7-iv.

Concluding that this suggests that hidden variables interpretations of quantum physics can not describe the quantum world is the same as making an erroneous assumption that these variables cause behaviour contrary to common sense; the exact opposite of the actual initial assumption hidden variables theorems make. In other words if we prove that hidden variables do not violate Bell's inequality we are simply proving that the two theorems are mutually exclusive. We are <u>not</u> proving that any one of the theories is wrong or right. Any such deduction is tantamount to saying that the Copenhagen interpretation is the absolute and unchallengeable truth handed to us by God himself. The logic might seem ok but this fundamental underlying assumption on which it is based is by all interpretations misleading.

All in all it would seem that there are no strange actions at a distance and there is no superposition of state. There are only hidden variables which we currently do not have the means to exhaustively determine.

7.4 Schrödinger's Cat

Einstein was not alone in trying to rationalise quantum physics in the 1930's. The deceptively scruffy looking bespectacled persona of Professor Erwin Schrödinger hid the intellect that dreamt up yet another famous hypothetical experiment in 1935. The express purpose of the '*Schrödinger's Cat*' thought experiment was again to reign in the young but wildly growing science of quantum physics by demonstrating the absurdity of Bohr's '*Copenhagen interpretation*'.

As if to complement Einstein's attack on Bohr's '*action at a distance*' implication through the EPR paper, Schrödinger singled out Bohr's '*Superposition of States*' idea. He found it difficult to accept the notion that an intelligent observer was **required** to '*collapse the wave function*' forcing a quantum system to assume a unique state. In other words that nothing exists until it is observed.

Schrödinger postulated an event which had a fifty-fifty chance of occurring such as the decay of a radioactive nucleus. The Copenhagen interpretation dictates that this nucleus will be in a state of superposition consisting of both the original and the decayed nucleus. Schrödinger suggested an experiment in which this superposed radioactive nucleus was placed in a box which contained a Geiger counter[41] which in turn was wired up to trip a switch that would release a cloud of poison gas into the box. A cat, hence the name of the experiment, would then be placed in the box and the lid closed. In strict observance of the Copenhagen interpretation not only was the nucleus in a superposed state but the Geiger counter was in both the state of having detected radiation from the nucleus and the state of having not detected this radiation. As a consequence the poison vial is in both the state of having released the poison and not yet having released the poison. The climax of this whole setup is the poor little cat which is said to be in both the state of a dead cat and a live cat. All this holds true according to the interpretation until someone opens the box and observes the set up at which point the nucleus collapses into one state or the other causing everything else to follow suit. This is a bizarre and illogical situation that prompted many young physicists to find alternative interpretations. This did not prevent Bohr's Copenhagen Interpretation for reigning more or less unchallenged until the 1990's.

---

[41] An instrument used to detect radiation.

The idea of superposition of states has its roots in Heisenberg's uncertainty principle (see 7.3.2). Because we do not know what state an electron is in, it seems logical to assume it is in both the possible states, a state of superposition. The Copenhagen interpretation simply served to formalise this concept. As explained before, the inability to determine the exact state of the electron is due to limitations in our perception of the world and not some weird dual character of the electron it's self. It's infinitely convenient however to brush this ineptitude under the carpet using something as simple as superposition of states (see 7.3.3).

7.5 The double slit experiment

Richard Phillips Feynman (1918-1988) is widely regarded as the father of modern day quantum physics. His indispensable contribution to the science from the 1940's to the 1960's was to place it on a secure logical foundation that naturally incorporated classical mechanics. Feynman had an uncanny, seemingly effortless ability to explain what seemed like complex scientific principles, clearly enough for non-scientist to understand. In a way he had an advantage over predecessors like Pauli, Heisenberg and Dirac in that he entered the arena of quantum physics in the 1940's when much of the ground work had already been done. He was spared working in the changing and complex environment of the 1920's and early 1930's. As a result he was able to channel his vast intellect more productively, sifting through the huge and varied amount of work done during these periods and logically building upon it.

In characteristic clarity through his lectures in physics published in the early 1960's he managed to distil the puzzlement around quantum physics to one basic phenomenon which seemed absolutely impossible to explain in any classical way. This phenomenon is the double slit experiment. Feynman went on to say that

'In reality, it contains the only mystery … the basic peculiarities of all quantum mechanics.'

7.5.1   The two faces of light

The history behind the double slit experiment goes back to the late 1600's and early 1700's. This age saw the birth of two competing theories on the nature of light, published respectively by the two greatest physicists of the time: Christian Huygens (1629-1695) and Sir Isaac Newton (1642-1727).

Huygens' book *'Treatise on Light'* published in 1690 put forward the wave theory of light[42]. This contribution was overshadowed at the time by the Juggernaut of a reputation wielded by Sir Isaac Newton who had published his history making *'Principia Mathematica'* 3 years earlier

---

[42] This theorised that light was made of ripples (waves) in an all pervading medium (the ether) which filled the entire universe. It's worthy of note that Huygens, in his book spoke of longitudinal (push-pull) waves and not the transverse (side-to-side) waves we use today.

in 1687. Newton went on to publish '*Opticks*' in 1704 in which he proposed the corpuscular theory[43] of light. The Newtonian theory of light saw no equal in physics until Thomas Young hit the scene one hundred years later and revived Huygens wave theory.

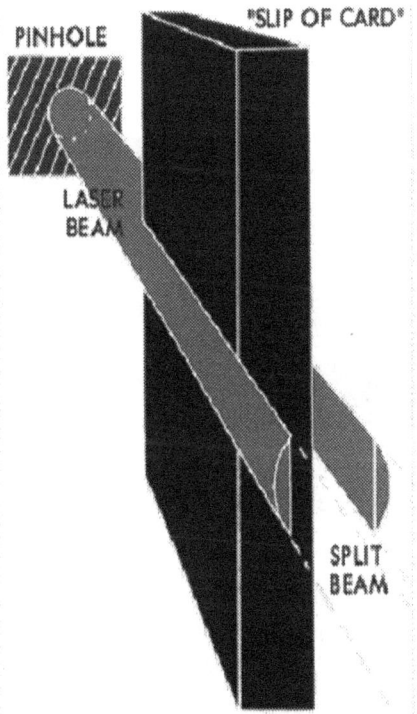

Figure 7-3 Young's original experiment setup

---

[43] This theorised that light was a stream of discrete particles, much like billiard balls. The principle argument behind it was that light cast sharp shadows rather than curving around obstacles like waves.

Though the famous double slit experiment is frequently attributed to Thomas Young, he in fact used a much simpler setup (see Figure 7-3) to demonstrate the wave properties of light. His setup involved a pin hole and a slip of card[44]. The resulting pattern (shown in Figure 7-4) could only be explained by the two split beams diffracting and spreading so as to interfere with each other much like the ripples in a pond diffract round an obstacle.

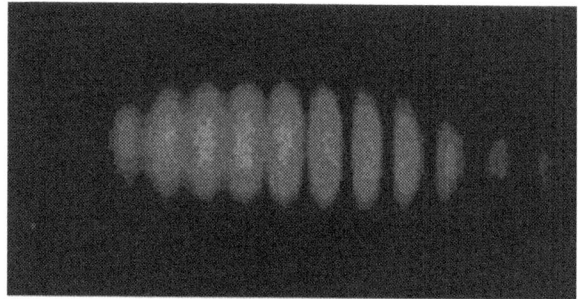

**Figure 7-4: The interference
pattern observed**

Even up to today the most telling experimental observation against Young's double slit experiment to argue for the corpuscular theory of light is that when light meets with an obstacle it creates a sharp shadow unlike waves which would bend into the shadow.

In 1909 Einstein was so convinced of the dual nature of light he is quoted as having said:

---

[44] More traditionally the modern day experiment is done using slits scratched in carbon deposited on a glass slide but the principle is the same

"It is my opinion, that the next phase in the development of theoretical physics will bring us a theory of light that can be interpreted as a kind of fusion of the wave and the emission theory"

Quantum theory's quantization of light into photons over the 30 years that followed provided a mathematical foundation for the corpuscular theory while also maintaining the wave theory. Even up to today we still refer to light as having a dual wave-particle nature.

7.5.2 Understanding the experimental results?

Mathematical physics of the early 1900's was predominantly Newtonian. From calculus to momentum and Universal gravitation, Sir Isaac Newton's enduring legacy was one on which many physicists felt confident to build and test their theories. It seemed only natural when the quantization of light in the 1920's was in agreement with Isaac Newton's corpuscular theory. Meanwhile, the experimental demonstration of light's wave characteristics could not be shrugged off.

Niels Bohr's vigorously promoted Copenhagen Interpretation had the daunting task of combining the successful corpuscular mathematics with the established wave centric observations in an attempt to '*explain*' quantum mechanics. Though it had its flaws, the interpretation stood unchallenged up until the 1980's and 1990's. This was the interpretation Feynman had to work with. Quantum theory had been born of mathematics which assumed that light traveled in discreet corpuscular units[45] know as photons yet accurate observations suggested that light was wave-like in nature.

When Feynman's contemporaries turned to the double slit experiment, they tried to physically realize the photonic particles so accurately described by their mathematics. It soon became apparent that even if one photon was sent towards the two slits an interference pattern was observed on the screen. They needed to explain how one photonic corpuscle could simultaneously pass through two independent slits and interact with it's self on the other end so as to form the observed interference pattern on the screen.

Scientist then fired an electron gun at the two slits. Logic had it that if light is particulate in nature like their equations so accurately predicted, then each electron fired will go through one and only one slit. It will **not** know that there exists another slit and over time each electron that goes through the apparatus will independently impinge on the screen behind. Surely there's no way an interference pattern would build up on the screen? Sure enough, impact after impact, dot by dot, gradually, the interference pattern formed. To make things

---

[45] Even though these equations, through Max Planck's efforts, did incorporate some wave parameters (e.g. $E = hf$ where f is the frequency of light) they considered light as corpuscular entities that just happened to have unexplained wave properties which could not be ignored within the mathematics.

even more perplexing, if a stubborn scientist tried to somehow detect which slit an electron or photon of light went through e.g. by covering one of the slits, the interference pattern would disappear.

7.5.3   The Quantum eraser

In one of the more recent spectacular variations of the double slit experiment popularized as the '*Quantum Eraser*', two polarizing filters are placed in front of the slits such that photons going through one filter are polarized differently to those going through the other filter (Figure 7-5).

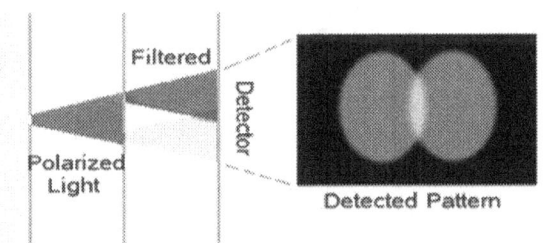

Figure 7-5 Double slit experiment with polarizing filters at each slit

We can now differentiate between photons passing through the two slits by measuring their polarization when they hit the screen. The interference pattern however fails to form upholding the quantum mystery. Then a scrambling filter is placed between the detector screen and the slits (shown in Figure 7-6) whose net result is to mix up the light such that it is no longer possible to determine which slit each photon passed through by measuring it's polarization like before. As if by magic, the interference pattern re-appears.

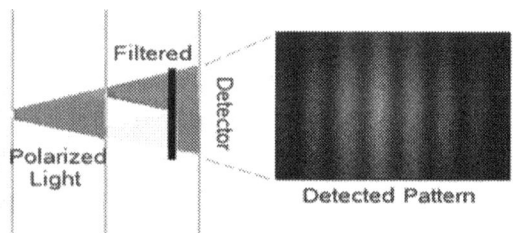

Figure 7-6 The Quantum eraser effect

The experiment was carried out using a beam of individual photons and the question remains as to how a single photon arriving at the slits knows to interfere when there is an eraser filter and not to interfere where the filter is removed. It hasn't even got to the filter yet at the point of interference. Is light a wave which collapses into particles when it is observed? It would seem that weird things are afoot in the quantum world. But are they? Is all of this confusion fundamentally because the mathematical equations are biased to the corpuscular camp while the physical observations are biased towards the wave theory camp forcing us to accept a weird and conflicting dual nature of light explanation?

7.5.4   Looking beyond the obvious

Physics and mathematics are deterministic sciences ultimately limited in their modelling of the world around us by our perception of reality. Many of the major advances in scientific physics however have not been initiated by an observation but by an idea in an adventurous young mind. Einstein's curvature of Space and time was such an idea. His paper on general relativity published in 1915 explained gravity yet no one had ever observed the curvature of space much less time. Even today scientists are still looking for ways in which to observe space-time curvature. One of the most ambitious of these attempts will be the LISA (Laser Interferometer Space Antenna). This will be launched in 2015 or later and will consist of three space crafts orbiting the Sun. They will use lasers to accurately monitor the distance between them in the hope that aberrations in this distance can be detected which imply the existence of the gravitational waves. This would confirm the predictions of Einstein's general theory of relativity. Again, up to today no one has ever observed curved space-time, if they did then surely they did not know that what they were observing was curved space-time or we would have heard about it. The mathematics however, has predicted many things that we do observe like the perihelion of Mercury. In this same light I'd like the reader to take a similar but logical leap of faith with me in an attempt to understand the double slit experiment.

Today we know that all matter including atoms has experimentally demonstrated wave properties.

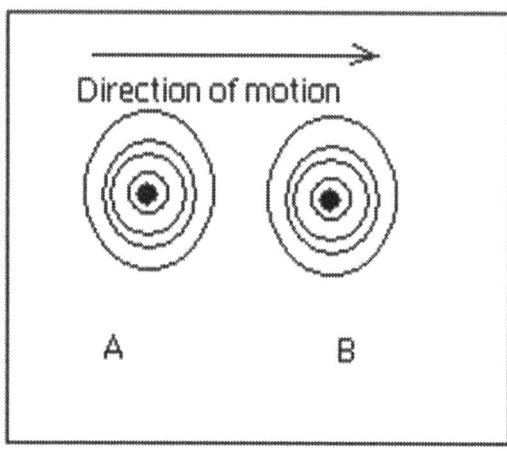

**Figure 7-7 Bird's eye view of two
vertically oscillating corks**

Imagine that photons are made of an infinitely small core of unknown constitution (maybe more concentrated light) whose dynamics result in a 3 dimensional wave-like perturbation of the space around them. These wave-like spheres of influence stretch out to a miniscule and equally imperceptible distance which is large in comparison to the radius of the central core. The analogy that springs to mind is that of atomic distances where it can be said that if the atom was the size of a football field it's nucleus would be the size of a pea. We can also postulate that this 3 dimensional wave-like perturbation of space influences the motion of the central core yet the very existence of the central core mandates the presence of the wave-like perturbation. There is an analogy here with Einstein's space-time curvature explanation of gravity which suggests that matter causes space to bend and space tells matter how to move.

Try to imagine what might happen when such a photon encounters the double slit experiment setup. It makes sense that the infinitely small central core will only pass through one slit if it is to make it to the screen. On the other hand, the wave-like perturbation of the space around it will spread out across both slits passing through to result in an interference pattern on the other side as expected.

The pattern generated on the screen can be explained depending on how we believe our eyes register light. If we assume that for us to see something we need the photonic cores described above to impinge on our retina then the bright strips of light at the screen are due to the incidence of the central cores of the photons described above. It then follows that the perturbation interference pattern between the screen and the slits must be acting as a guide for those cores that make it past the slits. On it's own it produces no image on the screen but by guiding photonic cores down gullies of constructive interference it concentrates them along strips on the screen.

Another explanation could consider that the moving islands of constructive interference that form between the slits and the screen become new photonic cores. In other words, the photon's surrounding perturbation pattern forms new photonic cores by constructively interfering on the other side of the slits. These, along with those that existed between the light source and the slits which were lucky enough to make it past the screen, form the familiar strip pattern on the screen.

If one of the slits is closed the photon generated wave perturbations will not be able to emerge from that slit so there will be no interference pattern on the other side. Those emerging from the lone open slit will spread out (diffract) causing the point of light to consume a greater expanse on the screen. Because there is no interference pattern from the surrounding perturbation waves the photon emerging out the other end will have no gullies to travel down. In fact they will be evenly distributed across the wave front in the maxima band and no interference pattern will be observed on the screen. Now imagine that different polarization filters at an angle to each other are placed at each slit. The waves through one slit will only be affected by the component of the polarized waves through the other slit that is effective in the direction of the subject wave's direction of polarization. If the angle between them is 90 degrees (the angle which enables us to unambiguously determine which slit a photon came through) then this component is effectively zero[46]. The smaller the angle the greater this component yet the less we are able to distinguish through which filter

---

[46] The component of polarised light in a direction at an angle θ to the direction of polarization is the value in the direction of polarization multiplied by cos θ. At 90 degrees cos θ is zero meaning there is no component at right angles to the direction of polarization.

and hence slit the photon passed. At certain angle the interacting components are maximum and great enough to produce an interference pattern however the angle required for this to occur is so small that the light from the two slits can no longer be distinguished by means of its polarization. In other words, any method that distinguishes between light from either slit implicitly prevents the waves through the slits from intermingling and hence producing the interference pattern.

But how about the eraser? Suppose the slit polarization filters are at 90 degrees to each other and no interference pattern is being observed as explained above. We then place a filter at 45 degrees to both these filters which consequently will let through the respective components from both slits at this angle. The result will have an equal contribution from both polarization streams but the light emerging from it will all be polarized at 45 degrees irrespective of the slit\filter it came through. It is now impossible to distinguish as to which slit a given photon passed through yet the components, now being in the same direction can interact with each other to produce the interference pattern. This third filter demonstrates the mysterious eraser effect.

In summary, by regarding photons as miniscule cores of intensely concentrated light (or even better, that which constitutes light) surrounded by wave like concentric, spherical perturbations of space it is possible to logically derive an explanation for the double slit experiment. Just like the curvature of space-time no one has ever seen this photonic structure, in fact, it begs the question that if we were to try to observe it what would we use. We can not use light photons since in our reality this is tantamount to perceiving the existence and detail of a 5 storey building by throwing another 5 storey building at it's general location and sifting through the debris. This technique is currently being used by the latest atomic colliders to understand the sub-atomic world but it is obvious that there is a lot of detail about the building that you will have lost in the resulting rubble which can only be perceived by using something that does not destroy the building. Mother Nature seems to prefer nuclear structures from the astronomical suns and solar systems orbiting the centers of galaxies to the electrons positioned round the atomic nucleus. The nuclear structure of photons therefore does not seem such a far fetched idea.

7.6 Discussion of quantum applications

7.6.1 Quantum computation

The essence behind quantum computation is the theory of superposition of states which as discussed earlier (see 7.4) is not a directly observed phenomenon. It is a theoretical inference from the implications posed by the limitations imposed on us as human beings when trying to perceive the entirety of the world around us. No one has ever observed a superposed quantum entity. In fact the theory dictates that this is impossible since observing the entity causes it to collapse into one state or the other.

Today's computers operate off one basic principle: on and off. They make decisions based on the binary state represented logically by a bit (short for binary digit) which can take on a value of 1 or 0. The computer combines these bits into 8 bit words (bytes) which can then be used in a complex description of the current state of the system as well as to communicate instructional and state information from one circuit in the computer to another. All this is orchestrated at great speeds and in such a small volume of space making it possible for the computer to perform a great many operations per unit time. In fact, the record as of March 2005 is held by IBM's Blue Jene at 135 teraflops (135,300,000,000,000 floating point operations per second). At the time this was expected to double to about 300 teraflops within 6 months. Just to appreciate how fast this is, the average novel contains about 100 words per page. Imagine each of these words is a random calculation to be evaluated like 1.423 + 5.68. Considering that the average novel is about 400 pages long you would need 6.5 million novels to write the number of such calculations that this computer can compute in a second. Imagine having to go through all of these books computing them all hand!

Quantum computation is being looked into at the moment as a way of exponentially increasing this capability. This will be achieved by replacing the bits with Qbits (Quantum bits). Unlike a bit which can only have one value (on or off) at a given instant a Qbit is a superposed quantum entity which according to theory is simultaneously in two or more states until it is observed. Now suppose you have a Qbit that represents 4 states superposed (i.e. the numbers 1, 2, 3, 4) then somehow perform a mathematical operation between it and another such Qbit (representing the numbers 6, 7, 8, 9) somehow resulting in a new Qbit being created which holds the results of the operation (i.e. 7, 9, 11, 13 if the operation was addition). If you can somehow read these multiple states without causing the bit to collapse as per theory you will have performed 4 mathematical operations instantaneously in theoretically the same time it takes to add one ordinary bit to another as is traditionally done by today's computers. If you then consider these Qbits placed in Qbytes and used in the same way as bits in computers you can theoretically exponentially ramp up the processing capabilities of today's computers. Is it really possible?

The first and most discouraging difficulty in achieving this is that it is based on a theory that actually prevents it from being achieved. According to the theory a quantum entity remains in its state of superposition until it is observed at which point it collapses into one of the possible states. The action of observing, measuring or in any way trying to determine the state of a quantum entity causes it to loose it's quantum weirdness so all you will ever get is one measured resultant state. We can not directly observe the constituent states of a superposed entity without disturbing it. The theory would have to be revised, hopefully as a result of further research and progress in the field, before a usable Qbit can be realised.

The other more practical difficulty is in finding a way to construct a Qbit stable enough to be used as part of a computational process. It's hard enough trying to understand what's going on within the quantum realm but it seems even harder to encapsulate this behaviour within a stable unit especially within the environment in which today's computers are expected to work.

Lastly but not least, Computers are traditionally deterministic machines. Yes they do perform myriad items of computational work per second but at any instant the computer as a whole (at the level of bits and bytes) is in a completely deterministic albeit complex state. The state might be too complex and dynamic to determine at that instant but it is deterministic all the same. Everything the computer does will depend on a set of deterministic rules and algorithms. Straight away it becomes obvious that the non-deterministic Qbit does not fit into this traditional architecture. As explained before (see 7.3.7) theory defines the quantum world as being ruled by statistical probability. It follows that the result of incorporating this into the deterministic computer we are accustomed to will require a total rethink of computer science as a whole. If I press the letter A on a computer, so long as all the logical pre-programmed paths are followed from the keyboard through the computer and to the screen where I am running a word processing application I expect the letter A to appear on screen. If it does not then it's about time I bought a new computer. Even the rarity of bit rot[47] is intolerable requiring error-detecting circuitry and at the very worst, machines and software are scraped or rebuilt.

Imagine writing and testing a computer program for such a machine governed by statistical probability. How would you ensure the program will behave as advertised to your clients? What will you do when it behaves unexpectedly knowing full well that this is the nature of the quantum domain you are working with so replacing the computer will not achieve anything? By it's very definition, a simple action like pressing the letter A should no longer be expected to result in an A showing up on the screen. The resulting state of the screen is now measured in terms of statistical probabilities. This nightmarish extreme scenario is not a prediction but a graphic illustration of the significant obstacles faced by quantum computation. At the moment, we are not even sure that this quantum computer is possible much less that it can be realised. But history has shown that it is when we are faced with seemingly insurmountable obstacles that

---

[47] Hypothetical disease in which unused programs or features stop working after sufficient time has passed, even if '*nothing has changed*'. The contents of a file or the code in a program become increasingly garbled over time by radiation

the best in human determination, ingenuity and innovation is born.

7.6.2   Quantum cryptography

Cryptography[48] has been a part of civilisation for centuries. Whether it be speaking in parables or using the "*Enigma*" machine, we have always had a need for private, uncompromised conversation, even more so in today's world where it is as easy to communicate with someone on the other side of the planet as it is to talk to your next door neighbour. The trend however has been that as technology advances so do the tools available to code breakers in there efforts to eavesdrop on encrypted communication. Quantum Cryptography offers an alternative solution to the modern conventional means of facilitating encrypted conversation.

To carry out an encrypted conversation the two parties (traditionally referred to as Alice and Bob) must solve two basic problems.

1. How to exchange a secret bit of information that will allow them, and only them, to decrypt messages from each other.

2. How to make it impossible for an eavesdropping third party (Eve, as she is traditionally referred to) to crack their encrypted conversation.

---

[48] The science of scrambling communication in order to hide its informational content and prevent its undetected modification or unauthorized use.

The asymmetric encryption protocol[49] solution to these problems is used widely in industry today but is under unrelenting pressure to keep ahead of technology and the hackers' persistent efforts to crack the code. Quantum cryptography offers an alternative solution which concentrates on the physics of the actual communication rather than the computationally vulnerable algorithms used by traditional methods. It allows Alice and Bob to communicate information and be able to detect if this information was compromised and how much so. This area of application of quantum theory has been so successful that in 2005, companies such as Magiq (www. magiqtech.com) started shipping practical, albeit costly, implementations. Surely such a practical application evidences the correctness of current quantum theory?

---

[49] In this Alice and Bob individually each generate two digital keys using an algorithm based on prime numbers. One of the keys is public and is sent to the other party respectively without any protection against eavesdroppers. The other key is kept private to each party respectively. The algorithm that generates the keys and that which uses them to encrypt messages is such that if a message is encrypted with Alice's public key, which Bob or anyone else may have, then only Alice can decrypt and understand this message using her private key. The possession of the public keys by the public domain exposes the conversation to attack however any brute force attempt to crack the conversation though not impossible requires more time and computing power than can be obtained. As technology advances, making this power available, the cryptographic industry needs to search for larger prime numbers or better algorithms to keep ahead of the hackers.

One possible setup is conceptually the same as that shown in Figure 7-2. Entangled photons are produced at the source and sent to the two observers (Alice and Bob) at A and B respectively. Once Alice and Bob have collected enough observation data with their polarisation filter and detector setup they should expect to be able to observe a correlation (see 7.3.6) between their respective data. This would violate Bell's inequality as expected. If this correlation is not observed then Alice and Bob know that someone must have intercepted their communication and thereby interfered with the data transmitted. They can then calculate how much of the information was compromised. Based on the results of this a decision can be made on whether to use the data to generate their respective secret keys or initiate a fresh communication for this purpose, scrapping the compromised data.

As explained in chapter 7.3.6 there are other more logical ways of explaining quantum observations than those used in the Copenhagen Interpretation. In addition, if we seriously consider the explanation given in chapter 7.5.4 we can arrive at an intuitive description of the physical process of quantum entanglement that makes this quantum cryptography application possible.

If two photonic cores are brought close enough such that their surrounding wave-like perturbation patterns overlap we will get a situation where the wave pattern generated by each photon exerts an influence on both the generating photon and the other photon. It is not difficult to imagine that certain properties will be commonly influenced across both photons through this overlapping perturbation pattern (remember, the existence of the pattern influences the motion and other characteristics of a photonic core within that pattern). When these entangled photons are separated the net effect of the earlier proximity is to **relate** certain of their properties in such a way that measuring them on one allows an inference of the value of the same properties on the other.

This is much like the situation that occurs when two entities at different temperatures are brought together and interact such that if you separate them again, knowing the resultant temperature of one of the entities enables you to make inferences on the resultant temperature of the other entity without measuring it. If a third party steps in and makes a measurement before you, they will undoubtedly affect the temperature of the measured entity no matter how careful they are (you can not measure something without interacting with it). If your measurements are accurate enough they will be able to detect this interception as a deviation from the expected resultant temperature and infer from this that someone else has interacted with the measured entity.

As you can see, quantum cryptography is not a proof of modern day quantum theory but an implementation of the statement that in order to make a measurement on an entity we have to somehow, directly or indirectly, interact with that entity. Whether this is via directly touching the entity or via bombarding the entity with detectable photons, the interaction will influence the entity being measured.

Now suppose Eve was endowed with or somehow discovered a new way of perceiving the world around her (other than light, see 6.2.3.1) such that she could make quantum observations with this new conduit of information with insignificant influence on the quantum entity she is measuring. Much like how you and I can observe a building without our photons punching a hole through it or causing it to collapse or change macroscopically in any way perceivable. It follows that she would be able to make observations on the photons being shared between Alice and Bob without affecting them in a way that the two can detect thereby successfully eves dropping on the key exchange procedure and compromising the whole subsequent cryptographic communication. With this in mind Quantum cryptography can only be practicable as long as we are all similarly limited in the accuracy with which we can perceive the world around us without influencing it.

7.6.3   Quantum teleportation

The phenomenon of entanglement has also nurtured ideas on which the concept of Quantum teleportation is based. This promises the ability to communicate information at speeds that make Einstein's universal '*speed of light*' limit seem like a snail's walk in the park. The more exotically inclined scientist actually see it, in theory, being used to instantaneously transfer individuals to the other side of the universe. How is this all possible?

The term teleportation is used in science fiction to refer to an object or person disintegrating at one location while a perfect replica appears at another location. When Captain Kirk is beamed from the star ship Enterprise to the surface of an extragalactic planet, he appears to de-materialise on the ship and re-materialises on the planet. Is captain Kirk broken down into his constituting atoms which are then physically sent down to the planet and reconstituted? Or, is it simply the case that all the information needed to recreate Captain Kirk is read from him in the Enterprise and sent down to the planet where it is used to construct him again using the raw materials on the planet's surface? How does quantum teleportation claim to have achieved this?

Let's start with our pair of entangled photons. One of the photons is given to Alice who flies to South America with it. The other is given to Bob who stays with it in London. According to the Copenhagen interpretation of quantum physics, when Alice opens the box containing her photon and observes a property on it, the photon held by Bob, thousands of miles away is affected. It instantaneously collapses into a state in which it exhibits a value for that property correlated to that observed by Alice. In other words, whatever interaction Alice has with her photon appears to have a correlated effect on Bob's photon (that being said this is not necessarily a correct inference to make, see 7.3.6).

Alice then gets hold of a third photon from South America and somehow entangles this photon with the one already entangled with Bob's photon. Implicitly, according to the theory, Bob's photon now starts getting information on the instantaneous state of the new South American photon picked up by Alice. Any observation made on this new photon will instantaneously have a correlated effect on Bob's photon. When Alice makes her measurement on this new photon she phones Bob telling him what measurement she made and the results. By using this to apply a transformation to his setup in London, Bob can make measurements which are correlated to and reflect the current state of the South American photon. He can then pass this information (via entanglement or any other means available) on to a photon locally in London. The net result is that the photon in London will be an exact replica of the South American Photon however, because it was observed, the quantum state of the South American photon will have been destroyed.

Let's take this a step further by recognising that we and all objects in the world around us constitute of quantum entities at the fundamental level. Suppose Alice gets hold of a willing, bikini clad South American beauty and splits her into her constituent quantum entities. One by one she passes them through this channel of communication with Bob reconstructing the beauty at the other end. No doubt that when the reconstruction is complete, Bob will have far more interesting things to do with his spare time than continue playing quantum ping pong with Alice.

As explained in chapter 7.3.6 this is all only possible under the Copenhagen Interpretation of Quantum physics which emerged from the confusion of the 1920's as the standard interpretation not because it had been proved beyond any reasonable doubt but because it was championed by the hugely influential person of Niels Bohr. Such was Bohr's reputation that his interpretation managed to withstand intellectual pressure from the likes of Einstein, Podolsky, Rosen and Schrödinger up until the late 1990's when a new generation of mainstream scientists started to take alternative interpretations seriously. As illustrated in chapter 7.3.2 and in the discussion on quantum cryptography (see 7.6.2) there does not need to be a spooky action at a distance to explain quantum entanglement and its consequences. Implicitly then, the current theory around the use of quantum behaviour to implement quantum teleportation can be explained in such a way that there is no instantaneous 'action at a distance'. This would mean that we are not observing teleportation but the macro effect of sub quantum properties which we can not directly observe. This is not to suggest that teleportation is indeed impossible. Knowing the limitless capabilities of human ingenuity, one day we might very well find a way of instantly being transported somewhere else in the universe. The current thinking around achieving this using quantum physics is however little more than a weird and wonderful hypothetical extrapolation of the already enigmatic quantum science into the realm of the unknown.

7.7 *Beyond the quantum*

Mathematics has been used from time immemorial by the human race to quantify the world around us. When we realised that there are aspects of this world which can not be quantified deterministically but on a probabilistic level, the branch of probabilistic statistics came into existence. Unfortunately every rose has its thorn. If I am to accept the fact that there is a one in two chance of a flipped coin landing showing heads and not tails, then I must accept that there is an infinitesimal chance that the coin will stop in mid air then for some reason move off into the corner of the room before falling to the floor. The probability might be so small that it is insignificant for all practical purposes however it is still a probability all the same and is an implicit requirement of the probabilistic mathematical model. We must appreciate in this light that when statistics is used to measure and try to understand the quantum world, it is simply a model of the underlying reality and not a direct point-for-point translation of this. That way we can objectively weigh all the inferences and extrapolations we go on to make from this representation of reality accordingly and appropriately.

In Dalton's time the atom was the smallest unit known to man of which all else is made. Today, leptons, which are significantly smaller than an atom, appear to hold this title. With this history it would take a braver person than myself to predict what our forays into that part of the jungle that is Quantum physics will uncover in the next decade or two. One thing is for certain though. We'll get to wherever we are going much faster if we all keep an open and objective mindset about ourselves.

8 Science, Today and Tomorrow

"Science is in a state of crisis. Where free inquiry, natural curiosity, open-minded discussion and consideration of new ideas should reign, a new orthodoxy has emerged."[50]

Logic is the foundation of science. In arriving at an understanding of the universe, logic prescribes a process of elimination starting from a set of all the available possible solutions and gradually trimming off those that fail to predict or naturally explain new observations or phenomena. So why do so many of today's respected scientist nurture pet theories which they adjust or even fudge to account for new observations such that they can wrongly claim that the theory naturally accounts for and predicts them, after the fact? Why do many of today's scientists evangelically preach that these constantly changing theories are unequalled models of reality?

---

[50] Rochus Boerner (2003) **"The Suppression of Inconvenient Facts in Physics"**

8.1 Power and Influence

Within the scientific institution, authority and influence is a tool with which to rebuke or even censure any non-conformant views. This is in stark contradiction with the principles on which science is traditionally believed to stand. Whether it be deciding which research to publish in prestigious journals or which investigative endeavour to recommend for funding or even how much funding resources to set aside for the research, usually the final decision lies with an elite who often harbour vested interests in favour of or against certain proposals. As João Magueijo and Andy Albrecht found out when their initial attempts to publish work on a brave new theory were rejected. Their Variable Speed of Light (VSL) theory attempts to correct problems with the Big Bang theory yet appears to challenge Einstein's universal speed of light limit[51]. They were told that in order to get published by *Nature* they...

> *"...would have to do more than show that our theory was a solution to the cosmological problems--- our theory would have to be the solution."*

It can be inferred from this statement that all solutions published in these journals are "*the*" solutions however many, however contradictory to each other and however much these are revised and fudged over time to account for

---

[51] "**Faster Than the Speed of Light**". Dr. João Magueijo, 2003

new observations. Even if they stood the test of time, can we as humanity (much less some independent reviewer for *Nature* magazine) really claim to know enough about the universe to identify "*the*" solution? Does such a solution even exists and even so, does it lie within our realm of perception? The two protagonists turned to "*Physical Review Diary*" hoping for a more objective and scientific criticism of their work but ended up in a year long review process that seemed to pitch institutionally arch rivals against each other taking the focus largely away from the scientific content of the submission. Today VSL theory is entertained by a growing number of scientists and has many papers published in its regard. As fate would have it, the irony is that Dr. Magueijo regards the whole pier review process rather cynically:

"*...as a chore not dissimilar to flushing the toilet or emptying the garbage bin.*"

A damning statement for the institution from someone within, who follows in the footsteps of the likes of Paul Dirac.

Ironically, it is the same power and influence which facilitates the adoption of brave new ideas however right or inaccurate they might be. Without this, Niels Bohr would have had a great difficulty popularising the Copenhagen Interpretation (see chapter 7.2.4) above the other interpretations of quantum theory that existed at such a time when so little was known for certain about the quantum world. This interpretation was not founded on irrefutable logical evidence but on the influential persona of Bohr and it is only in the last decade or so that it has been found to be less than satisfactory in many respects as a representation of the quantum world. Scientist are now beginning to seriously consider the alternative hidden variable interpretation that had been up to now unfairly ignored and discredited.

The scientific institution does not seem to give due credit to Heresy and dissidence. It is easy to forget that much of the foundations of the institution were built on the efforts of anti-establishment individuals. Galileo was imprisoned for not conforming to accepted beliefs and Einstein himself was relegated to the confines of a patent clerk's office arguably for his academic non-conformance. In fact, there is a striking similarity between the institution's leading authorities, who edit and review scientific journals and the heads of the catholic inquisition of the 15th century. Thankfully it is no longer the heretic's life that is at stake though the amount of damage they can do to his or her career is quite a deterrent. As if that is not enough to sink the ship, enter those individuals and organisations whose wealth and clout manages to redirect efforts towards their vested interests rather than the unadulterated pursuit of knowledge. How has humanity arrived at this disheartening set of circumstances?

8.2 Laws imposed on Nature

Part of the problem is that the institution is built on shaky foundations which by there very nature are weakened even further with each new contribution through the insistence that this must conform to the existing foundations. Those that laid the founding stones did not have the foresight that in a few hundred years these would have to support the innumerable implications, theories and new observations that we have today. Consequently, any inaccuracies in these founding principles have been magnified over time as the institution grew, placing it in a position where either it imposes its self on the contributors and their contributions in a desperate bid to stay alive or it is demolished and replaced with stronger, more contemporary foundations which though they too are still prone to inaccuracies will provide a better platform from which to develop our further understanding of the universe. In fact, Humanity will be in abetter position in a few hundred years with this more solid foundation whose principles are more rigorously tested and explicitly qualified to protect against misuse and misinterpretation. So what are these shaky foundations?

8.2.1 The misdefinition of energy

Energy as a concept percolates through all of physical science as we know it today in one way or another. It would seem like without energy nothing is possible whether it be to fly a man to the moon, to collide particles together or simply to walk down the street, everything requires energy. Energy is defined as the ability to do work a definition which presents a double edged sword. It both facilitates the concept to be easily woven into the entirety of physical science and at the same time exposes the concept to varied interpretations and applications which build on top of each other to arrive at concepts which though aesthetically beautiful no longer seem to naturally fit into the reality described by the original concept.

8.2.2   Why is E=MC² Einstein?

In the fourth of his ground breaking 1905 papers[52] Einstein discussed his theory of special relativity. At the end of the paper, almost as if it were meant as an afterthought, he proposed the now famous equation $E=MC^2$. What might have been overlooked as a footnote to the casual reader of his paper can today be credited (or cursed depending on which of the myriad ways you look at it) for ushering in the atomic age. Einstein did later formally publish a mathematical proof for this equation[53] giving it the firm footing within the scientific establishment that it continues to enjoy even today. It therefore might come as a surprise that a discussion around this central pillar of modern science actually illuminates glaring and shocking cracks within the scientific process and institution.

---

[52] **"Does the Inertia of a Body Depend Upon Its Energy Content?"** Albert Einstein (1905).

[53] **"Relativity: The Special and General Theory"**, Albert Einstein (1961)

There are numerous formal proofs for E=MC² out there many of which are quite lengthy and not for the faint hearted however, just like all the fundamental equations of physics it can be logically arrived at from first principles with minimal mathematics. Our starting point lies at what we understand to be the fundamental definition of the very subject of Einstein's equation. Energy!

One well known quantitative definition of energy is that the work done (energy expended) in moving an object through a given distance is equivalent to the force applied (f) multiplied by the distance through which the object is moved (d). This is sometimes referred to in principle as the "*Work Function*"

$$E = f \times d \dots\dots\dots\dots \text{8-i}$$

The equation describes Energy in Joules (J) in terms of Force in Newton's (N) and distance in meters (m). It gained its popularity during the age of the steam engine. Then, it was used to great effect when calculating how much fuel had to be burned in order to move a given load over a given distance. This simple equation has since served engineers well and consequently earned its way into high school physics text books. But, what happens if the force is applied to a stationary object? That interesting point shall be discussed later on but for now, let's consider the other classical equation to be used in our proof.

$$f = m \times a \dots\dots\dots\dots \text{8-ii}$$

This equation states that when a given constant force (f) is applied to a constant mass (m) which is free to move, the mass will experience acceleration (a).

If we substitute for (f) in equation 8-i we obtain equation 8-iii below.

$$\textbf{E} = \textbf{m} \times \textbf{a} \times \textbf{d} \ldots\ldots\ldots\ldots \textbf{8-iii}$$

In other words, an amount of energy (E) will be required to give a constant mass (m) a uniform acceleration (a) as it travels a distance (d).

For simplicity let us assume that the body of mass (m) starts its motion of uniform acceleration (a) from rest. At the point when it has completed traveling the distance (d) it's instantaneous velocity will be v_d. In this case the uniform acceleration (a) will be given by.

a = change in velocity / time taken for change

$$= (v_d - 0)/t$$

$$= v_d/t_d \ldots\ldots\ldots\ldots \textbf{8-iv}$$

Note that V_d is the instantaneous velocity at which the mass (m) is travelling at when it has travelled a distance (d) and (t_d) is the time it has taken to cover this distance (d). If we now use equation 8-iv to substitute for acceleration in equation 8-iii we get

$$\textbf{E} = \textbf{m} \times v_d/t_d \times \textbf{d} \ldots\ldots\ldots\ldots \textbf{8-v}$$

We can shuffle the terms around in this equation to obtain something that resembles our objective as shown in equation 8-vi

$$E = m \times v_d \times d/t_d \dots\dots\dots\dots 8\text{-vi}$$

Remember, (v_d) is the instantaneous velocity exactly at the point when the mass (m) has covered the distance (d). But what does the term d/t_d mean? This term is the ratio of the distance travelled by the mass during this experiment and the time taken to cover this distance. It is represented as v_m in equation 8-vii.

$$E = m \times v_d \times v_m \dots\dots\dots\dots 8\text{-vii}$$

This is a generic equation so let us see what happens at the instant when the mass attains the speed of light. At that point the equation becomes:

$$E = m \times c \times v_m \dots\dots\dots\dots 8\text{-viii}$$

Note that in this case v_d (as originally defined) is equivalent to c (the speed of light) at this instant yet v_m is a totally different quantity. In fact, if we dare to state that v_m is equal to c we shall be effectively implying that a mass (m) moving at the speed of light for a time (t) will have covered the same distance (d) as an identical mass (m) starting from rest and accelerating to a final velocity equal to the speed of light in the same time (t). If that's a bit of a mouthful then a more palatable analogy is due.

Athletes in a 100 meters race start at rest when the whistle blows then accelerate to a blistering speed in the first 50 meters or so which they more or less maintain for the final 50 meters. We spectators watch in amazement as they cover the distance at times well within 10 seconds. Now, imagine that Carl Lewis (one of the fastest men ever over this distance) challenges his clone (who we shall refer to as Carl Lewis X) to a race. To make things a bit more difficult for himself, Mr. Lewis allows Mr. Lewis X to start 50 meters before the starting line and promises that he will wait at this line and only start running when Mr. Lewis X actually crosses this line. From then on it is a 100 meters race to the finish line. Assuming none of them get tired during the race, while Carl Lewis spends the first 50 meters accelerating to his cruising speed, Carl Lewis X is already at this speed throughout this 50 meters. After Mr. Lewis attains the cruising speed the two athletes will be running at the same speed however Carl Lewis X will be ahead having been traveling at this maximum speed for the entire race. Substituting v_m for c in equation 8-vii is tantamount to stating that Carl Lewis will overtake his clone Carl Lewis X within the first 50 meters even though he starts from rest and never actually attains the speed of Lewis X until after the end of the said 50 meters. In fact, we are saying that Mr. Lewis will inexplicably end up actually winning the race. Under the assumption that the two athletes perform identically this can never happen within the framework of physics and reality that we know today.

Logic dictates that v_m must be less than c if the mass at a constant velocity and the identical accelerating mass are to cover the same distance in the same time. I will refrain from going into the details here but it can be mathematically proved that v_m will always be substantially less than c. Consequently $E=MC^2$ will always describe an amount of energy greater than the energy of the mass as defined by classical physics. So long as there is no experiment conducted that can claim to measure and account for every last iota of energy contained within a given mass (providing that there is no energy lost to the surroundings and unaccounted for), there will be no way to disprove Einstein's equation through experiment.

A stationary object also has inertia but in this case it is proportional to the mass of the object. It can be said that as a body moves faster and faster it gains in inertial mass (this is separate from the body's actual mass which does not change). The force required to decelerate the body is equivalent to the force required to similarly accelerate another body at the same rate who's mass is equivalent to the inertial mass gained. In other words, the energy expended in moving the body ever faster seems to be creating a 'conceptual' mass which gets heavier as the body moves faster and faster. It is this conceptual mass that the equation is referring to and not the constant mass of the moving body. However, once this conceptual mass becomes equal to the mass of the moving body we can logically conclude that the energy expended in getting to this point is equivalent to the energy inherent within the constitution of the moving mass.

8.2.3   The stationary body and $E=MC^2$

Though Einstein only considered bodies in motion it is important to note that the same conclusion above can be arrived at when we consider circumstances which are not catered for by equation 8-i. This equation which is the very basis of the text book definition of the concept of energy only takes into account moving bodies. It does not account for the energy expended when a force acts on an immoveable object. It is understandable why during the age of the steam engines there was little reason to popularise an equation which effectively works out how much energy is expended while performing no useful work (seeing as the immoveable object does not undergo any visible or useful change in nature or position).

If you apply a force to an immoveable lamp post you will eventually get tired in proportion to both the magnitude of the force you apply and the length of time for which it is applied. Suppose that two individuals apply the same force independently to two separate immovable objects for the same period of time. It follows logically that they must have expended the same energy irrespective of the material of which the objects are made. In other words, even though the energy expended is transferred to potential energy and dissipated into other forms like heat differently for the different materials, the total energy input is proportional to the force applied (f) and the time for which that force is applied (t). It is tempting to write this in terms of equation (8-ix) below but we realise that doing so will contradict the long revered equation (8-i).

$$E = f \times t \ \text{................} \ 8\text{-ix}$$

If we juxtapose both equations 8-i and 8-ix we find that they are identical except that one talks of time (t) where the other talks of distance (d). This would imply that time (t) is fundamentally equivalent to distance (d). Such an assertion would wreak havoc within the 'aesthetic beauty' of modern day mathematical physics. That said however, Einstein's theory of general relativity seemed to suggest that the two are one in the same thing[54].

The two physical quantities are measured in totally different and seemingly unrelated units. It is however interesting to note that for a body whose motion is constant we can predict how far it will have travelled in a given time and implicitly therefore, the distance moved is related to and can be used to measure time. The time elapsed is also related to and can be used to determine the distance. In other words, for such a body, distance and time are directly convertible.

Now, in the contradicting scenario above, one of the equations has to give and it can not be the time tested equation (8-i). Traditionally, whenever new equations are proposed which appear to describe reality yet contradict existing, time tested equations in this way scientists apply a scaling correction which is later referred to as a constant. The units of this correction are deemed such that the new equation no longer contradicts with existing equations thus

---

[54] There is an analogy here with Einstein's concept of space-time where space (the realm within which distance exists) and time are brought together under the concept of the space-time continuum thus simplifying the form of most physical laws.

everyone is happy[55]. If we avoid upsetting the establishment and keep to this convention equation (8-ix) can be written as:

$$E = k \times f \times t \text{ .... } 8\text{-}x$$

Where k is a scaling constant of yet unknown value whose units are $JN^{-1}S^{-1}$ (Joules per Newton per Second). It follows that multiplying this (k) by the force in Newton and the time in seconds will yield a value who's units are Joules (J) which are the units of energy we already know as described in equation 8-i. The magnitude of (k) can now be determined through quantitative empirical observations of the variation of E with the product $f \times t$[56].

Now, consider an experiment where a certain amount of energy E_1 is expended in moving a force F_1 through a distance d_1 as per equation (8-i). We then set up another experiment in which the same force F_1 is applied to an immoveable object and we somehow monitor the amount of energy being expended in the process. As soon as the total energy expended reaches the same magnitude as the previously noted E_1, the experiment is stopped and the elapsed time t_2

---

[55] Another more detailed illustration of this referring to Newton's gravitational Constant can be found in "The Final Theory", Mark McCutcheon (p27 – p48).

[56] Note again here that we can not determine the total quantity E absolutely since a lot of the energy is lost immeasurably to the surroundings so we can not really prove this equation until future generations solve this energy loss problem or establish a different model of the concept of energy which addresses the issue.

is noted. E_1 and F_1 are common and unchanging factors in both experiments. If we place these factors to the left of equations 8-i and 8-x we get a situation where:

$$E_1/F_1 = d_1 \quad \text{and} \quad E_1/F_1 = kt_2 \text{ .... 8-xi}$$

This implies that:

$$d_1 = kt_2 \text{ .... 8-xii}$$

In other words the distance travelled by a moving body when a force F expends energy E to move it is directly proportional to the time taken for the same force F to expend the same energy E on an immoveable body. Notice that we are assuming that we can measure all the energy expended in the two experiments absolutely and without energy loss. The implication again is that an element of distance can be converted into an element of time. Upon successfully conducting this experiment over and over again we can arrive at a value for (k) which will be equivalent to the ratio of d_1 to t_2.

k=d1/t2

The units of (k) are hence the same as the units of velocity and for all intents and purposes (k) does indeed represent a particular speed[57] which I will term V_1.

Now, from equation (iii) we can substitute for k as follows

---

[57] I have not carried out this experiment nor heard of anyone that has but it would be interesting to note and compare the value of k obtained to the known speed of light.

$$E = d_1 / t_2 \times f_1 \times t_2 \dots\dots\dots\dots \text{ 8-xiii}$$

Assuming that the moving body starts from rest

$E = d_1/t_2 \ m \ d_1/t_1{}^2 \ t_2$

$E = d_1{}^2/t_1{}^2 \ m$

$E = mv_1{}^2$

Now say that the final velocity of the moving body is c then the energy contained by that body will be given by:

$E = mc^2$. So a body moving at or close to the speed of light embodies this amount of energy.

It is interesting to note that the smaller the particles we can perceive are the closer they get to the speed of light and that our modern day model of the constituent of matter considers these very small particles as the fundamental particles from which visible matter is made. Why is it that all of them travel at a speed less than that of light. Could it be because we can not perceive anything faster than the speed of light due to the limitations discussed in chapter 6.2.3 that it appears to us that this is a physical speed limit whereas it is just a boundary between the world we can perceive and that which we are still unable to perceive. The smaller things get the faster they are until we approach the speed of light then we seem to not be able to see or perceive anything at or beyond this speed. We interpret this by saying that nothing can travel faster than this speed but forget that light being our conduit of information has now become inadequate to make any concrete conclusions about reality. Just as when something approaches the speed of sound we can no longer use sound to perceive its existence (see chapter 6.2.3.2). This implies that if we had another means of perceiving the world (e.g. gravity, which appears to act almost instantaneously on all bodies) we would find that the speed of light is not a speed limit at all. Though gravity affects us and gives us a sense of direction (i.e. up and down) we merely know of it's presence but do not have a sense of gravity (the ability to perceive and differentiate entities in reality through gravity)

8.2.4   The Graviton and the theory of everything

The graviton is a hypothetical particle through which gravity is realised in Standard Theory. By all accounts this particle must travel faster than the speed of light to account for the instantaneous nature of gravity. Given the trend that the smaller things are the faster they get, it is logical to assume that this particle must also be very small. Note that these characteristics combine to make its detection next to impossible with today's technology. Not only do you have to find a means of observing something moving faster than the speed of light but you also have to get this miniscule thing to interact with the matter that our senses can sense and affect it such that we can infer its existence. This is like not being able to observe a tennis ball but depending on the effect it has on the earth when it is thrown towards it and to make matters worse, imagine that this earth is pervious to tennis balls. We can't observe it directly because it is travelling faster than the fastest means we have of perceiving reality and we can't observe it indirectly because it does not seem to have an individual effect on objects we can observe which is perceptible. Instead the effect we observe is the communal effect of gravity from which we can make limited conclusions and assumptions about the individual particle.

Even if we could perceive a graviton, would this be the theory of everything or simply the theory of everything we know and can perceive about the universe. Historically, the trend has been that there is always something we need to know to understand and relate existing theories but having this new knowledge opens up a whole new world, giving birth to whole new theories which takes us back to where we were in the first place. When Dalton described atoms as billiard ball like indivisible particles the scientific world thought they finally had the theory of everything by knowing that everything consists of these indivisible atoms. All the then known scientific theories could be explained in terms of these atoms. Of course as they got to know the atom better they found that it is not really indivisible and new theories like quantum mechanics were thought up to replace the Newtonian mechanics that had previously sufficed. These new theories have opened up a whole new world and now scientist a searching for a way to reconcile it with the beautiful Newtonian physics of old. The so called Theory of Everything. When Dalton devised his atomic theory he did not think it mandatory to reconcile this with the earth, wind and fire philosophy that reined a few hundred years earlier. His genius was in recognising that this was a new, unrelated and superior means of describing and modelling the world.

8.2.5 The speed of light limit (why can't we accelerate or decelerate light)

As discussed in chapter 6.2.3 the speed of light being a physical speed limit is more of an imposition of law on nature rather than a characteristic of absolute reality.

According to Einstein as a body accelerates it gains in inertia and it also gains in kinetic energy. Einstein realised that towards the speed of light the body seemed to have an infinite inertia (or mass) and with such it would take an infinite force to change the motion of this moving mass whether it be to accelerate it towards the speed of light or to decelerate it. This force increases as the object approaches c. Einstein argued that therefore nothing can produce that little extra bit of acceleration to get the body to and beyond the speed of light.

Nothing we know of can do this but what if some cosmological force exists which we do not know of much less understand which is responsible for accelerating the infinitely small gravitons to superluminal speeds. This would explain the instantaneous action of gravity but because they are very small and travelling faster than light we can not perceive them therefore we assume they do not exist.

What Einstein should have said is that as far as our perception of the world around us is concerned nothing travels faster than the speed of light because light is the fastest conduit of information we have to perceive the world. Much like the man on the bus described in chapter 6.2.3.2 we are imposing our limitations in perceiving reality on reality its self in a futile effort to render absolute reality as consisting of only that which we can observe. Throughout history, humanity has always been reluctant to accept that there is much more to the universe and reality than that which we are aware of because this renders us next to insignificant in the grand scheme of things.

All these misdefined or misunderstood concepts[58] are just a few out of many. To aggravate matters, scientists build entire careers out of these concepts and thus have cause to fight tooth over nail against any attempt to logically argue against them as this would render all their life effort having been wasted. This is why many of them were shocked when Stephen Hawking recently stepped out and doing the institutionally unthinkable, said, "*I was wrong*

---

[58] For completeness another example is the struggle over the last few decades by veteran Astronomer Halton Arp to get the institution to accept new observations made which prove that Hubble's expansion is a myth and hence the concept of redshift is misunderstood and misdefined. A full account is offered in his book "**Seeing Red: redshifts, Cosmology and Academic Science**".

about black holes." Nobody is used to respected figures revising or even dumping their strongly and evangelistically held beliefs.

8.3 Conclusion

The final judge of science should be the entirety of humanity but the institution has thrown a dark fog of complex models over our heads to keep us occupied and out of there way while they commandeer the ship as they see fit. It would seem that they are the only ones gifted enough or capable to pursue the cause. It's a dictatorship of sorts and such governances will rarely end peacefully. We need a revolution to get back what rightly belongs to the whole of humanity and not the select elite amongst us.

We have a choice, we can accept without argument, the anthropocentric definition of the world as presented to us by Einstein, Bohr, Hubble and others. In this case reality and the world around us is absolutely defined by the way we perceive it and by implication, nothing exists beyond our limits of perception i.e. beyond the speed of light. Though it is logically weak, this argument leaves us within the comfortable illusion of significance in a deterministic universe. Alternatively, we can accept that our perception of the universe is only an infinitesimally small portion of the absolute reality of the universe which bears well logically but leaves us as insignificant entities in a wilderness which we seem to have little hope of conquering. Perhaps the greatest irony of all is the following quote from Albert Einstein himself:

> "Unthinking respect for authority is the greatest enemy of truth."